W0234435

Science for Sustainable Societies

Scope of the Series

This series aims to provide timely coverage of results of research conducted in accordance with the principles of sustainability science to address impediments to achieving sustainable societies – that is, societies that are low carbon emitters, that live in harmony with nature, and that promote the recycling and re-use of natural resources. Books in the series also address innovative means of advancing sustainability science itself in the development of both research and education models.

The overall goal of the series is to contribute to the development of sustainability science and to its promotion at research institutions worldwide, with a view to furthering knowledge and overcoming the limitations of traditional discipline-based research to address complex problems that afflict humanity and now seem intractable.

Books published in this series will be solicited from scholars working across academic disciplines to address challenges to sustainable development in all areas of human endeavors.

This is an official book series of the Integrated Research System for Sustainability Science (IR3S) of the University of Tokyo.

More information about this series at http://www.springer.com/series/11884

Hiroshi Komiyama • Koichi Yamada

New Vision 2050

A Platinum Society

 Springer Open

Hiroshi Komiyama
Mitsubishi Research Institute, Inc.
Chiyoda-ku, Tokyo, Japan

Koichi Yamada
Center for Low Carbon Society Strategy
Japan Science and Technology Agency
Chiyoda-ku, Tokyo, Japan

ISSN 2197-7348   ISSN 2197-7356   (electronic)
Science for Sustainable Societies
ISBN 978-4-431-56622-9   ISBN 978-4-431-56623-6   (eBook)
https://doi.org/10.1007/978-4-431-56623-6

Library of Congress Control Number: 2018936906

Printed on acid-free paper

This Springer imprint is published by the registered company Springer Japan KK part of Springer
Nature.
The registered company address is: Shiroyama Trust Tower, 4-3-1 Toranomon, Minato-ku, Tokyo 105-
6005, Japan

Preface: From "Vision 2050" to "New Vision 2050"

During the twentieth century, energy-supported material civilization advanced significantly. This brought about resource depletion and climate change. How should material and energy be utilized in future to bring about global sustainability? In 1999, "Vision 2050" (Komiyama) was depicted as a model that should be pursued half a century into the future.

Twenty years have elapsed since the first description of Vision 2050. Fortunately, the world is moving toward Vision 2050.

The twenty-first century will probably be a century in which we will seek qualitative affluence with Vision 2050 as a materialistic base. This lends itself to the specific image of a platinum society that is replete with resource self-sufficiency, coexistence in harmony with nature, life-long self-reliance, diverse options, and free participation. This book presents a "New Vision 2050," to which the viewpoint of a platinum society has been added to Vision 2050.

The concept of "New Vision 2050" is described in this Preface and Chap. 1.

Preface (1): Turning Point of Human History

Twenty-First Century Is a Special Era

Figure 1 shows the global average per capita GDP, life expectancy, and CO_2 concentration levels from the year 1000 AD to the present. The three lines depict similar trajectories that rise rapidly upon entering the twentieth century. Given the pace of expansion of human activity, the twentieth century stood out as a very special era.

For the most part of their long history, human beings have developed very slowly. During the Greek and Roman eras, humans lived to about 24–25 years old, and even after the passage of more than 2000 years in 1900, the average life expectancy was still only 31 years old, an increase of only a few years. However, in 2011, the average life expectancy came to exceed 70 years. The greatest reason for this was the gains in material wealth—a marked change from the past, where people lacked

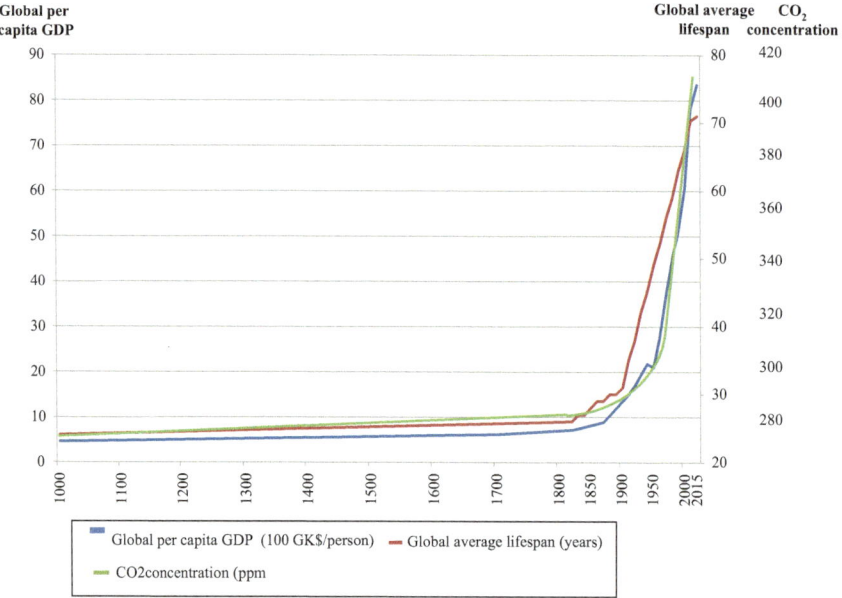

Fig. 1 Trajectory of Sudden Expansion in Human Development in the 20th Century. (Source: Per capita GDP and average life expectancy created based on Angus Maddison, while CO_2 concentration created based on NOAA)

clothing, housing, and food. As a consequence, however, human beings now face various problems such as an aging society and have started to alter the earth itself, resulting in such changes like global warming.

We are living in a special era of human history.

Various Issues

A wide range of issues confronts modern society, including aging society, environmental issues, low-growth economy, energy, resources, disparity, unemployment, terrorism, coups, fiscal issues, pension, infectious diseases, obesity, and lifestyle-related diseases. As indicated in Fig. 1, it is thought that the root causes of many of these issues are a reflection of an era of explosive expansion of human activities. In other words, the development of civilization itself appears to be responsible for these issues.

While some people suffer from hunger and malnutrition in some parts of the world, the amount of food waste continues to increase, and excessive eating and obesity have become social problems in developed countries. As reflected in Fig. 1, the twenty-first century is a very special era in the sense that it encompasses a broad spectrum of issues. There are people in the early development stage as well as those who have already exceeded that stage, highlighting issues ranging from slave labor to excessive eating and obesity, as well as environmental challenges.

Global Warming and Abnormal Weather

Climate change is one of the biggest threats. The average global temperature has continued to rise, just as scientists predicted. With torrential rains occurring frequently in Tokyo, typhoons making landfalls in the Tohoku region, and snow falling in Hawaii, the phrase "once in 100 years" has become a common truth. Will the ecosystem be destroyed? Can agriculture, fisheries, and the tourism industries cope with the situation? We do not yet have answer to these questions.

Understandably, many people worry about the impact of changes in the earth, which is the foundation of the survival of humanity (Fig. 2).

Fig. 2 Damage caused by a super typhoon
In Chuuk, more than 6,500 people have been displaced from their homes because of the damage caused by Typhoon Maysak
By Erin Magee, USAID/OFDA (https://www.usaid.gov/crisis/micronesia)
Public domain via Wikimedia Commons

Is Capitalism Sustainable?

Concerns have arisen over civilization based on capitalism and democracy. In the twentieth century, the world saw two ideologies, namely capitalism and communism, competing for supremacy. In 1991, the Soviet Union collapsed, and excitement filled the air over the end of the Cold War. However, this has been short-lived, given that regional conflicts centering on developing countries continue to occur frequently, and the fear of terrorism has decidedly enveloped the world. Aggravating the situation, the UK's decision to withdraw from the EU (Brexit) has poured cold water over advances made throughout history and has raised concerns that nationalism and ultra-nationalism have become rampant throughout the world.

One important explanatory variable is the widening economic disparity. There is no doubt that in low-growth developed countries, the wealth gap has widened between groups that have capital and are earning profits that exceed the economic growth rate, and groups have no such capital and whose wages have been falling. In addition, the gap between developing countries and developed countries is still wide, and together with the disparity within developed countries, these disparities have been increasingly destabilizing for the world. Can human society really continue to be sustained?

The sustainability of civilization found in capitalism, and in turn based on democracy, has been called into question (Fig. 3)

Fig. 3 Refugees fleeing from their homeland in smuggling boats
LE Eithne rescuing migrants as part of Operation Triton (2015). By Irish Defense Forces. https://commons.wikimedia.org/wiki/File:LE_Eithne_Operation_Triton.jpg
Licensed under https://creativecommons.org/licenses/by/2.0/

Preface (2): The Latest Report

Human Race Is Navigating in the Right Direction (1): SDGs

There is hope. At the United Nations Sustainable Development Summit held in September 2015, an agenda for "Sustainable Development Goals (SDGs)" consisting of 17 goals was unanimously adopted (Fig. 4). The human race navigated in the right direction at one of the most important turning points since the dawn of history.

The SDGs were preceded by the "Millennium Development Goals (MDGs)" formulated in 2000, which consist of eight goals including "the eradication of extreme poverty and hunger," "reduction of the infant mortality rate," and "improvement of maternal health." Certain results such as halving the extreme poverty rate were achieved by the promised year of 2015. However, there remain people who suffer from poverty in the world. There are also infants who are not properly vaccinated. Childbirth in the absence of doctors or midwives is still not uncommon.

Against this backdrop, the SDGs were formulated as a "plan of action for people, the Earth, and prosperity" by expanding the target audience to all humans and taking over developing countries issues addressed by the MDGs. To be more comprehensive, the number of goals increased from eight to seventeen.

The United Nations proclaimed that although the world faces a multitude of challenges, "no one will be left behind."

Fig. 4 Logo of SDGs created by the United Nations. (Source: United Nations Information Center)

Human Race Is Navigating in the Right Direction (2): COP21

At the 21st Conference of the Parties (COP 21) to the United Nations Framework Convention on Climate Change held in Paris, France, on December 13, 2015, the Paris Agreement was unanimously adopted by all 196 participating countries and territories. Once again, mankind made the right choice on the brink of history. The success of COP 21 provided hope for humanity that we can believe in.

The Paris Agreement called for limiting global warming to an increase of no more than 2°C as a universal long-term goal with 1.5°C as the ambitious goal. The goal is to eventually achieve zero-carbon emissions status (Fig. 5).

An increase of 2°C would already have a significant impact. Abnormal weather would occur frequently, the melting of polar ice would accelerate, and the primary industry would be severely damaged. Contrary to initial beliefs of global warming, which held that the impact of global warming was limited only to some areas such as Tuvalu, a small Polynesian island in the Pacific, its impact is in actual fact more far-reaching than that. Countless environmental catastrophes such as the typhoon that struck the Philippines, and tornadoes and torrential rainfalls that have started to strike various parts of Japan, are all believed to be caused by climate change.

For many years, scholars maintained a conservative standpoint that climate change, as understood from a long-term perspective, and sudden weather abnormalities are two unrelated concepts. However, they have since begun to ascertain that much of the observed weather phenomena is clearly abnormal. The opinions of these scholars, coupled with the observed environmental impacts on humans, likely led to the Paris Agreement.

The human race has finally acknowledged that climate change is as much an established fact as it is a threat, and has agreed to overcome it.

Fig. 5 Crowd at venue jubilant over adoption of COP21 Paris Agreement. (Courtesy: WWF Japan)

IEA Report

The Paris Agreement is not merely a hope. According to the International Energy Agency (IEA), global carbon emissions have remained constant at 32.1 billion tons during the four year period from 2013 to 2016. It is possible that CO_2 emissions, though the overall level continues to increase, may have finally hit a ceiling. Over the past three years, global GDP has achieved a growth rate exceeding 3%. It is a commonly held belief that economic growth results in increased energy consumption and an increased level of CO_2.

However, the 3-year period from 2013 revealed that economic growth does not necessarily result in an increased level of CO_2. This is referred to as the decoupling of the economy and energy. The world has henceforth entered an era of simultaneously aiming at economic growth and suppressing CO_2.

The main reason that CO_2 emissions have hit the ceiling is that such emissions have started to decline in China and the United States, which are the first and second largest CO_2 emitters in the world, respectively. In China, the consumption of coal decreased due to increased energy-saving tactics, and at the same time, use of renewable energy sources that do not generate CO_2, such as hydroelectric and wind power generation, has increased. Power supply from renewable energy sources increased from 19% in 2011 to 28% in 2015. In the United States, the utilization of renewable energy has increased, alongside a shift from coal to gas.

Therefore, it is possible to reduce CO_2 while achieving economic growth.

Japan's Experiences as a Leading Country in Resolving Societal Problems

If everyone can live life with dignity and material needs are satisfied, many difficulties can be overcome. For civilization to be sustained, it is necessary to secure people's material needs while maintaining the global environment. After all, we will be faced with the difficult task of simultaneously reducing energy consumption while achieving economic growth. In fact, Japan, as a problem-solving developed country, has the first-hand experience of dealing these problems and has shown that it is possible to solve them.

Figure 6 shows Japan's GDP, energy consumption, and power consumption over the past half a century. For two distinct periods of history, Japan achieved economic growth without any change to energy consumption. The first instance was during a 12-year period from 1973 to 1985 when its GDP increased from ¥200 trillion to ¥300 trillion while energy consumption remained unchanged. The second instance was in the most recent decade, in which energy consumption decreased while economic growth, although small, occurred. Electric power consumption also decreased.

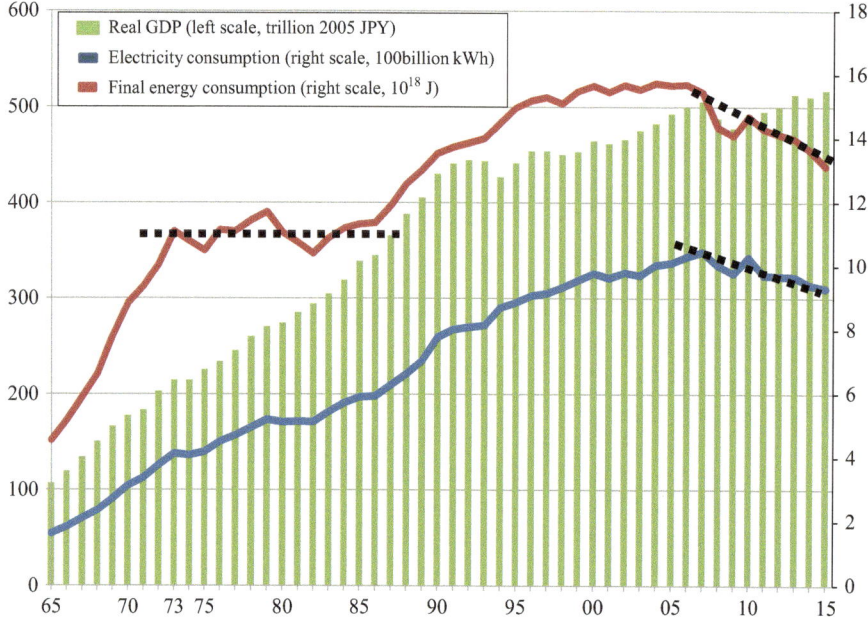

Fig. 6 Trajectory of Japan: GDP/Energy consumption/Power consumption. (Source: Created by the authors based on various materials. Real GDP: "System of National Accounts," Cabinet Office; Electricity demand and the final energy consumption: "Comprehensive Energy Statistics," Agency for Natural Resources and Energy)

Japan has led the world in decoupling economic growth and energy consumption. The factors that caused this decoupling were industrial energy conservation during the first period, and more recently, energy conservation and saturated demand for man-made objects.

In this way, the experiences of Japan as a leading country in resolving societal problems have given credence to the viability of the direction that the world should pursue. The fact that global CO_2 emissions have hit a ceiling since 2013 suggests that this model of decoupling can occur in the world.

Essence of the Era of Saturation

Satellite photos have provided us a compelling image of Earth floating in space. The fact that the Earth's temperature, which is a basic condition for our survival, can change because of human activity reveals that the Earth is but a small planet of limited size and resources. In this finite Earth, human activities cannot expand indefinitely. Saturation would thus be a fitting term to describe the twenty-first century.

Saturation of the Population

Will population explosion lead to the destruction of mankind? This belief has been popularized until half a century ago, although many experts continue to express such an opinion today. However, it is probably just baseless fear.

Figure 7 shows the number of births in the world. If the number of children born reaches a point of saturation, the overall population is likely to become saturated. Likewise, if the number of births decreases, the overall population would likely decline eventually. Although the future population depends on the number of births and average life expectancy, the number of births is of foremost importance in the population forecast, given that average life expectancy is headed toward a plateau. Looking at the figure, the number of births that once peaked in 1988 has begun to increase again from 1998. This can be attributed to the increase in the number of births in Africa. However, outside of Africa, the population is projected to decrease.

The number of children a woman can bear usually decreases when the GDP increases. History has furthermore shown that the number of births decreases when women are educated. This is clear from the current situation in developed countries where birth rates have declined, resulting in a struggle to maintain their populations. For the future, it is important to suppress population growth by promoting education in Africa. If this is carried out properly, the fear of population explosion will subside.

Rather than worry about population explosion, we must shift toward expanding efforts to maintain the population. Many developing countries have done so, not to mention developed countries.

The twenty-first century is a century in which the population becomes saturated.

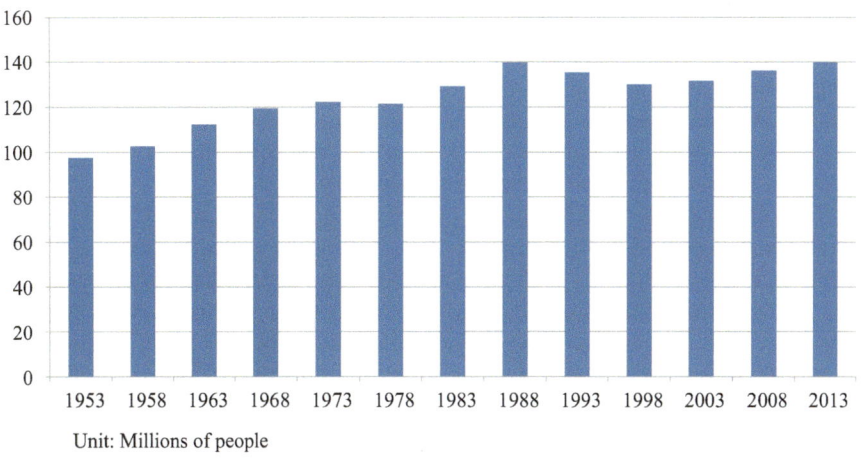

Unit: Millions of people

Fig. 7 Changes in the Annual Number of Live Births in the World. (Source: World Population Prospects, 2015 Revised Edition)

Saturation of Man-Made Objects

Table Intro-1 shows the ratio of automobile ownership to the population for each country. A ratio of 0.5 indicates that one car is owned by every two persons, a level at which car ownership reaches a saturation point, regardless of the country. When saturation occurs, an equilibrium state is reached where the number of new cars sold becomes equal to the number of cars discarded. In Japan, this number would be 5 million cars, which is derived at by dividing 60 million cars by 12, the average number of years it takes for a new car to be scrapped. In other words, domestic demand for automobiles is saturated.

In developed countries where automobiles are saturated, the consumption of electrical appliances, buildings, plastics, as well as many other manufactured goods has also become saturated. Looking at the process of development of a country's economy, first, infrastructure such as roads and railways is constructed followed by production of basic goods, such as fertilizers and fibers. As people start to become affluent, they acquire electric appliances, etc., and as they reach the peak of affluence, they start to demand automobiles. Therefore, in countries where automobiles are saturated, other man-made objects are also saturated. In developed countries, the sales volume of man-made objects has reached saturation point. This essentially explains the low growth rate of advanced countries.

It is a matter of course that "things will become saturated in this finite Earth", but perhaps it is also matter of course that we tend to overlook this obvious fact.

Saturation of Minerals

Iron, produced from iron ore, is the most important material in terms of both quality and quantity. It is produced from iron ore. Once made, iron will never be discarded. When buildings are dismantled, iron materials such as steel frames are recovered as scrap, melted, and supplied to society once again. Therefore, when the number of man-made objects saturates, and the amount produced equals the amount discarded, the iron supplied by breaking down the discarded man-made objects will be sufficient to sustainably make all the new man-made objects.

Figure 8 shows the total amount of iron contained in man-made objects such as buildings and automobiles that currently exist in Japan as well as their annual increments and decrements. The total amount of iron has approached about 14 billion tons, and the amount accumulated every year has been approaching zero. In fact, the amount of iron used in man-made objects in Japan every year and the amount of scrap iron have become equal at about 30 million tons respectively. Therefore, in Japan's case, the iron supply from recovered scrap is sufficient, and hence it is no

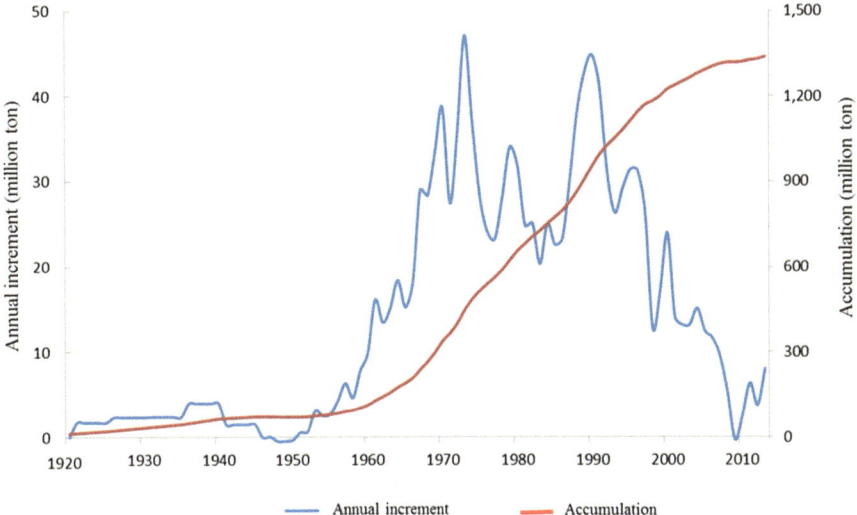

Fig. 8 Total Accumulated Amount of Iron in Japan and Annual Amount Accumulated
(Data) The Japan ferrous raw materials association

longer necessary to produce iron from iron ore. It turns out that this is not unique to
Japan; many developed countries no longer require iron ore. Nearly all regions of
the world will likely approach this state around 2050.

Let's generally refer to this state of saturation seen in automobiles and iron as the
"saturation of man-made objects." From the viewpoint of the current resource-
mining-based economy, the "saturation of man-made objects" will likely cause
negative effects such as drops in the price of raw resources and decrease in economic growth. On the other hand, however, this indicates that mankind is headed
toward a recycling-oriented economic society. The twenty-first century is an era
marking the shift to a recycling-oriented society due to the saturation of man-made
objects.

World in 2050

At this present time, the human population, man-made objects, and minerals are
already becoming saturated in developed countries. The rest of the world is also
catching up rapidly. Around the time that those who are high school students in
2016 turn 50 years old, the world will have reached an era of saturation. This is not
merely the reality of a distant future.

Preface (3): A Society We Are Aiming At

Creation of a New Society and Values by Resolving Issues

In the twentieth century, people were released from various constraints. Many people became free of constraints such as labor for securing food, clothing, and shelter; had inability to move; and are not being able to acquire information. The dream of longevity has also been realized.

As we acquire these things, it may not be that easy for us to recognize that we "have in fact become free." Consider the predicament of farmers of the Edo period. Lacking food, clothing, shelter, and without access to external information, they relied on traveling by foot and manual power, and endured hunger when struck by crop failure. Compared to their situation, today's reality is like a dream. Even if we do not go back to the Edo period, we were forced until recently to choose between working in the more urban Pacific Ocean belt zone and remaining in the countryside. Now, the situation has changed completely. Although it is not always evident, we can now freely choose where we want to live or work. In this sense, we have become free.

Even if we cannot deny that there are negative aspects of our civilization, we have fulfilled our longstanding dream of affluence and longevity. A question that we might ask now is whether we humans have the wisdom and ability to take advantage of the freedom we have gained, and work toward creating a better society.

However, it is unlikely that we will have to think in a new, difficult manner. Rather than resolving the issues that need to be resolved under current restraints, we should resolve them by expanding on the freedoms we have gained. By doing so, a better society will naturally emerge; one that will generate a new sense of values. Perhaps, we should proceed based on this thinking. Let's do that and move forward. We can participate in society by freely choosing from among various options such as hobbies and work, nature and stimulation, and friends and family. Such a society has come within our reach with our efforts.

"Platinum Society": A Vision in the Twenty-First Century

During the Edo period, for example, when food was in shortage, it was not necessary to come up with a vision. As people wanted to be able to eat, they had no choice but to diligently engage in farming. During the age of industrialization, people wanted to buy televisions or cars, and for that, they hoped for better salaries. In other words, becoming affluent must have been the implicit vision.

The vision in the twenty-first century must be one of a high-quality society. To be precise, it must be a society where people can maintain quantitative affluence, enhance it if necessary, and enjoy a better-quality life and living situation. Or in other words, a society where they can enjoy a better quality of lie (QOL). Let's define such a society as a platinum society. It is a society with a spread of brilliant colors such as eco (green), health (silver), and IT (scarlet). To achieve that, instead of denying the current situation that civilization has brought about, we should think about achieving a dramatic improvement of our QOL by adapting the present society and industrial structure to a platinum society. The image of a global community that we should aim for is one in which everyone, and not just those in developed countries on this Earth, is living in a platinum society. As declared in the SDGs, we should "leave no one behind."

Essential Factors for a Platinum Society

If we want to improve our QOL, we must first need to understand the present status of our QOL. In the first place, individuals are free to decide what quality is. Therefore, a platinum society is a diverse society in which individuals, communities, or organizations can make choices freely. However, there must be common qualities on which most people agree.

Here are some qualities on which many people agree:

1. No anxieties about resources and energy
2. No pollution and the maintenance of the global environment
3. Symbiosis (coexisting) with diverse and beautiful nature
4. Long-term health and self-reliance to be achieved for a long time
5. Opportunities for lifetime social participation
6. Ability to continue lifetime development
7. Employment opportunities
8. Cultural and material affluence

As can be deduced, matters as resource self-sufficiency, low-carbon, overcoming pollution and living in harmony with nature, health and self-reliance, lifelong development, diverse options, and free participation are essential qualities for QOL (Fig. 9).

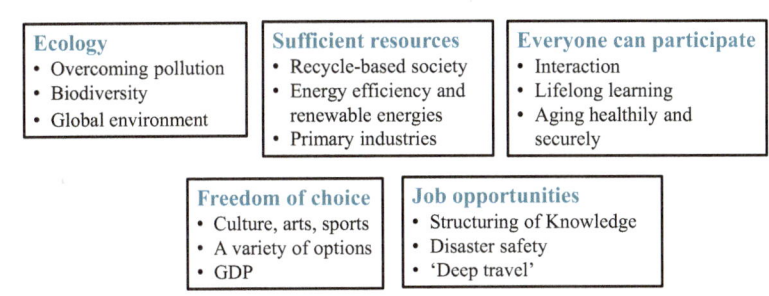

Fig. 9 Necessary conditions for a Platinum Society

With the transition to a platinum society, the way we work, the way we study, and our lifestyles are bound to change. These changes will bring increased freedom and business opportunities for people. They will also benefit from a sustainable society in which they will gain material and spiritual affluence. Because it will be a society without anxiety, people can have dreams for the future. Consequently, many will consider having a family and bringing up children. The answer to the declining birth rate problem, which has already become the most serious issue in many countries, is the realization of a platinum society.

A Vision That Can be Realized

Is a platinum society a distant ideal that is far from reality? This should not be the case. It is an ideal image that can be achieved. In fact, there are no small number of examples in which it has been already realized, albeit partially. The Platinum Society Network holds the Platinum Grand Prize Awards every year, presenting awards for activities aimed at realizing a platinum society (Fig. 10). These reflect a partial image of a platinum society.

Considering the greatness of the human strength that we can be demonstrated in these successful case examples, together with Vision 2050 in which the Earth will provide material affluence for everyone, I am convinced that a platinum society can be realized.

Fig. 10 Grand Prize, Platinum Vision Award

Preface (4): Image of a Platinum Society Has Begun to Appear

Creative Demand

When man-made objects become saturated, the saturated demand converges to a constant value. New demand is seen around new desires, or in other words, around the necessary conditions for a platinum society. We shall call that creative demand. Let's look at the image of the platinum society that has started to appear as well as new businesses to be spun off from it.

Low-Carbon Society

Figure 11 shows energy consumption in Japan. About 60% of the total energy consumption in Japan occurs in daily life activities, while the remaining 40% is used in monozukuri (making things). The breakdown of daily life activities is as follows: households (20%), offices (26%), and transportation sector (17%). Therefore, the key to the realization of a low-carbon society is to figure out how to reduce this energy consumption without reducing the quality of life.

Energy-Creating Houses and Zero-Energy Buildings

Homes will likely become places for producing energy rather than for consuming it. These are energy-creating houses where the amount of power generated by solar cells is greater than the amount consumed. A recent survey has shown that good

Fig. 11 Energy is being used by us. (Japan)

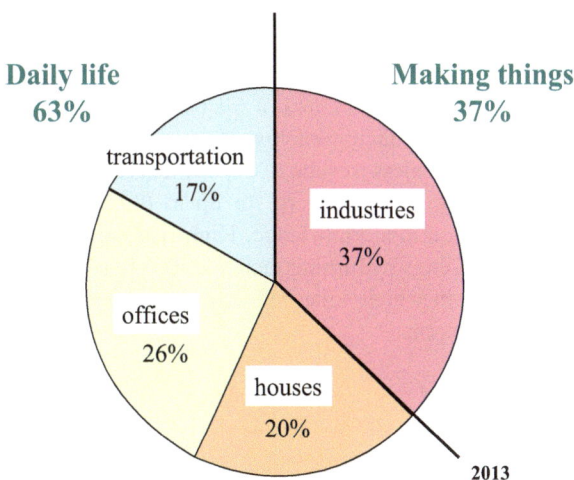

Fig. 12 Era of "solar-cell roofs". (Courtesy: Kaneka Solartech Corporation)

houses contribute to promoting the health of the people who live there. Houses will become the hub for comfortable living, promoting good health, generating energy, and contributing as an industry in itself. Fig. 12 shows an example of recent solar cells, with the solar cells serving as roofing material, and its elegant design. We are moving past the era of "putting solar cells" on the roof and entering an era where "the roofs are themselves solar cells."

Houses in Japan have been built to spend the summer in comfort. Therefore, while they provide good ventilation and their roofs look great, thermal insulation is very poor. The energy used in heating and cooling will diminish if thermal insulation is improved. At the same time, the energy efficiency of home appliances with high energy consumption such as refrigerators, air conditioners, and lighting has significantly improved. In newly built homes, reducing the energy consumption to about one-third has already become the norm. Energy-creating houses that provide power to the third party by turning the whole roof into solar cells are also already commercially available.

Cooling, heating, and lighting are at the center of energy consumption in the office. Yet the amount of consumption in the office is not that very different from in the home. However, the density of energy consumption is high, and in the case of high-rise buildings, the area of the roof where solar cells are installed is relatively small, and so reducing energy consumption to zero is more difficult than in the home. Nevertheless, results of experiments have shown that it is possible to reduce energy consumption to zero for up to three-story buildings, and this is already becoming common knowledge. High-rise buildings will also likely become zero-energy buildings by turning their "walls into solar cells" (Fig. 13).

The era in which solar cells on tile roofs of temples are commonplace is just around the corner.

Fig. 13 Case example of a zero-energy building (ZEB). (Courtesy: Taisei Corporation)

From Eco-Cars, Eco-Factories, and Cars to Cars, and Zero CO_2

Automobiles are at the center of energy consumption in the transportation sector. Electric vehicles and hydrogen fuel cell cars will likely become the mainstream before 2050. As we bid farewell to the era of cars running on gasoline, it will become a matter of common sense to not emit CO_2 while driving. Since these cars are highly efficient, CO_2 emissions will be drastically reduced even if electricity or hydrogen power will have to be produced from fossil resources. Eventually, electricity and hydrogen will be produced from renewable energy, and CO_2 emissions in the transportation sector will decrease to zero.

The "Environmental Challenge 2050" announced by Toyota Motor Corporation in October 2015 is indeed the automobile version of "Vision 2050" which I presented in 1999 (see Chap. 1). The world's greatest automobile company has turned—steered—itself in the direction where it should be heading.

Diversifying Means of Transportation

Transportation methods will likely diversify. For example, we may see young people riding bicycles and senior citizens driving mini electric vehicles (such as electric wheelchairs) on a sunny day. As products become more affordable, they will own

Fig. 14 Various means of transportation. (Courtesy of Platinum Society Network)

two types of such vehicles—one for use in good weather and one for rainy weather—and this may spur economic growth. Currently, experiments are underway to fly electric aircraft from a suburb of London to a suburb in Paris, and these planes are scheduled to become commercially available in due time. Currently, traveling between provincial cities is troublesome. These convenient electric aircraft could resolve that problem and make living in rural areas far more convenient (Fig. 14).

The freedom of information and movement has reduced disparities between regions substantially. However, although access to information has become easier, traveling on airplanes, Shinkansen bullet trains, and the like is still expensive. If the costs are reduced to 1/10 or even 1/3, people should be able to feel a greater sense of freedom. As seen from the entry of LCCs (low-cost carriers) which have made airfares cheaper, price collapse will probably occur due to innovation in the transportation sector. Innovation takes place where there is a need. One positive aspect of globalization is that innovations can occur anywhere in the world and can quickly spread throughout the world. In consideration of the above, we will likely become more affluent and gain more freedom as a result.

Energy Conservation Is the Best Policy

Significant progress will be made to promote the low-carbonization of daily life activities, i.e., households, offices, and the transportation sector, which together account for 60% of energy consumption. Energy conservation in the monozukuri industries, which account for the remaining 40%, has been steadily progressing, and will also continue to progress in the future. As I have asserted in Vision 2050, tripling energy efficiency from the level in 1990 is an achievable target.

Among businesses that promote energy conservation is the ESCO (Energy Service Company) model. Currently, the model is aimed at large companies, large buildings, and other large infrastructure. Potential future targets of energy conservation include small and medium-sized buildings, as well as homes, since there exist great opportunities for the ESCO business in those targets. Given the expected emergence of venture companies, they are also potential targets.

Urban Mines

Currently, approximately 70% of the world's steel is produced with iron ore as its raw material, while the remaining 30% is produced from scrap, which is waste material resulting from man-made objects. Enough amount of scrap is generated to produce necessary amount of steel in developed countries. Around 2050, a situation where the saturation of man-made objects and the generation of sufficient amounts of scrap-metal necessary for iron production will likely develop in many countries in the world. In fact, producing steel from scrap-metal utilizes only one-third of the energy required to produce steel from iron ore. The production of iron accounts for about one-fifth of the total energy consumed in the monozukuri industry. Therefore, urban mines will not only eliminate the need for iron ore but will dramatically reduce energy consumption.

This situation is not limited to iron. It also applies to copper, tin, zinc, rare earth metals, as well as precious metals such as gold. To turn urban mines into a more efficient system, it is necessary to recover man-made objects efficiently. Systems to recover buildings, houses, automobiles, and large home appliances to serve as scrap are already in place in Japan. Potential items for recovery will be small home appliances such as personal computers and smartphones, especially since about half of the precious metals and rare earth metals are contained in these items. While effi-

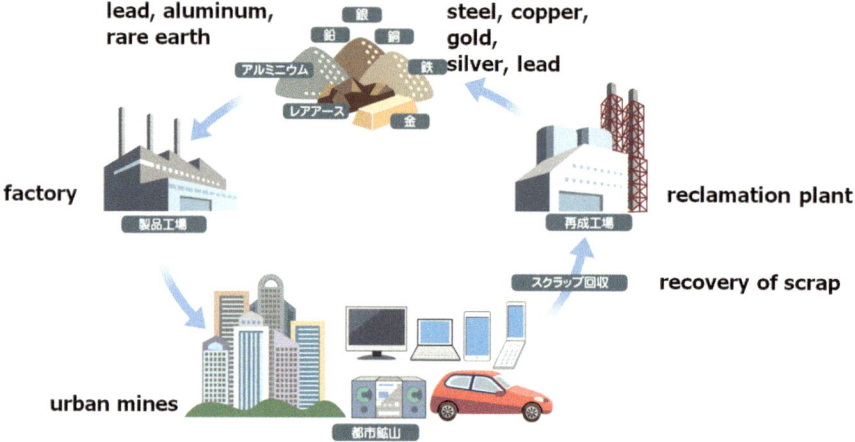

Fig. 15 Conceptual image of urban mines. (Courtesy: Platinum Society Network)

cient recovery of small appliances requires innovation, Lynette Japan has been taking on the challenge by collaborating with a home delivery company.

By around 2050, mankind will be bidding farewell to the era of metal mining, and transitioning to a society where metals excavated in the past will be recycled (Fig. 15).

Renewable Energy

A situation in which renewable energy sources such as hydro, wind, and solar power have become the most economical choices for building new power plants has begun to emerge in various parts of the world (Fig. 16). Emerging from an era of safe and secure but expensive energy, we now find that renewable energy can compete with fossil resources and nuclear power on price. Since innovations will occur in the renewable energy sector, the price competitiveness of renewable energy is likely to further increase in the future.

All renewable energies, except for geothermal and tidal, are a form of solar energy. The amount of energy from sunlight that hits the Earth is 10,000 times more than we need. Since it is practically infinite and the cost to harvest its energy has already become the cheapest among energy sources, there is no doubt that renewable energy will become the energy of the future.

In fact, it is likely that Japan will enter an era where renewable energy will be the norm before 2030.

The Internet has replaced centralized information systems, and both convenience and robustness have increased. As a clean, safe, robust, cheap, abundant, sustainable, and therefore, secure energy source, renewable energy will become the core of the energy system in the future, projected to occur by 2050.

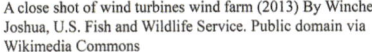

A close shot of wind turbines wind farm (2013) By Winchell Joshua, U.S. Fish and Wildlife Service. Public domain via Wikimedia Commons

Solar Panels (2014). By Michael Schwarzenberger Licensed under Public Domain C0.

Fig. 16 Massive wind power generation and solar power generation facilities

Energy conservation will advance, metal resources will be channeled to urban mines, and renewable energy will become the core of the energy source during the twenty-first century. This is not merely a dream of the distant future.

Japan Will Become a Resource Self-Sufficient Society

Renewable energy will replace fossil resources, and urban mines will replace natural mines. The result is a resource self-sufficient society. Japan will turn toward achieving self-sufficiency in resources from the traditional import and process of resources and export as finished products, which has been the national policy since the Meiji era. With people beginning to live in affluence although packed together in a limited geographic area, Japan provides us with an early image of Earth in 2050. Therefore, Japan's journey to achieving resource self-sufficiency as a problem-solving developed country presents a vision of the future of our Earth for the future.

The World Will Become a Sustainable Recycling-Oriented Society

If Japan's resource self-sufficiency is extended to the world, the world will become sustainable in terms of energy and other key resources. Specifically, this means a recycling-oriented society based on energy conservation, renewable energy, and urban mines.

Harmony with Nature

Figure 17 compares the conditions of the sky and sea in Kitakyushu City in the 1960s and the present. Various parts of Japan, especially around the Pacific Ocean belt zone, experienced terrible pollution in the past. Fortunately, nature has regained its beauty, and with its newfound status as an environmentally-advanced global city, Kitakyushu City has since been visited by many foreign delegations. Its experiences have come to be known as the Kitakyushu model and have inspired urban planning in more than 60 countries in Asia.

The main objective is not only about regaining beauty. Nature has taken over to the extent that wildlife such as storks and the Japanese crested ibis have returned to Toyooka City and Sado, respectively. Schools of ayu (Japanese trout) can also be found swimming upstream in rivers in Tokyo such as the Tama-gawa River.

Figure 18 reveals the transformations that the Genbe-gawa River in Mishima City have undergone. The river is located about a 10 minutes' walk from a station that is 40 min from Tokyo by Shinkansen bullet train. Visitors can enjoy watching fireflies there, and it is no surprise that tourists visiting Mishima City have doubled to 7 million over the past decade. As a result, previously vacant stores and shuttered (ghost town) shopping districts have disappeared. Mishima City is a successful example of how to strike a balance between an economy and society in harmony with nature.

Fig. 17 The beautiful sky and the sea that have been revived by overcoming the pollution of the 1960s. (Provided by Kitakyushu City)

1950s   **1980s**

now

Fig. 18 The number of tourists has doubled since clear streams over which fireflies dance have returned to the Genbe River in Mishima City. (Courtesy: Groundwork Mishima)

Upon seeing many good case studies and considering logical possibilities, I have come to believe that the trilemma of the environment, stable provision of material and energy, and economic growth can be overcome.

Macro-Level Viewpoint of Harmony with Nature

Japanese company Ito En has consolidated abandoned farmlands in order to produce tea leaves as raw materials for its products. The total consolidated land area is close to 400 hectares, and abandoned farmlands have been restored into beautiful tea plantations, as can be seen in the photograph (Fig. 19). For the company, ensuring the safe and stable supply of tea leaves has become possible by involving itself directly in the production. In the regions in which Ito En operates, agriculture has not only become a competitive industry but also revived employment. Its initiative has thus simultaneously resolved both business-related and social issues.

Farmlands and forests occupy 80% of Japan's land. Currently, about 7% of farmlands are abandoned, and 70% of forests are substantially left untended. It is thus

Abandoned mulberry field   Current state (Kitsuki district in Oita
                                   Prefecture)

Fig. 19 Abandoned farmland and a tea plantation that has been revived. (Courtesy: Ito En)

essential that initiatives are created from the viewpoint of the national land conservation. Ito En's experience is instructive in indicating the direction of corporate-led solutions.

Furthermore, mankind will soon face food shortages. Maintaining a stable supply of wood is also a serious challenge. If Japan fully engages in forestry, it can achieve self-sufficiency in wood, forests will be preserved, and mountains will be maintained. Maintenance of forests in southern regions will become easier, being relieved from the burden of overcutting in order to meet the export demand from Japan.

It is necessary to maintain a systematic perspective when examining the trees, forests, Japan, and the world at large.

Health Support and Self-Reliance Support Are Important Industries

There are concerns that longevity would lead to a stagnation of the economy. Such concerns are not unfounded, as it cannot be denied that elderly individuals do indeed possess less motivation for consumption than to young people. However, they have a stronger desire for health and self-reliance compared to young people. Demand emerges where there are needs, and if there is supply, economic activities will emerge. Therefore, new industries should emerge in the aging society.

The structure of medical services in the past was that for diseases, there were clear causes such as pathogenic bacteria and viruses, and drugs would be used to combat those causes. However, many people today are plagued by symptoms that

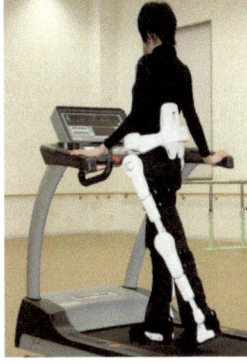

It is possible to live independently as long as the brain is alive.

Robot suit HAL
(lower-limb type, for care)
Produced by CYBERDYNE
(Japan)

Communication robot "Giraffe"
Produced by Robotdalen (Sweden)

Eating aid "Bestic"
Produced by Robotdalen (Sweden)

Fig. 20 Support for independent living is an important industry. (Courtesy: Mitsubishi Research Institute)

do not fit into these categories, such as lower back pain, atopy, and depression. The deciding factors, then, are big data and analysis, as well as the utilization of information technology such as artificial intelligence. It is not hard to imagine that a huge industry will be created from such technologies.

Figure 20 shows a robot that helps handicapped persons. When people think or act, brain cells emit electrical pulses. There exist a huge number of electrical circuits in which those impulses flow through nerves, reach the point of action and return to the brain cells. This is how people perform actions. Therefore, if the robot activates a motor by detecting a current to determine a person's intentions, it will become a robot that supports humans. Electrical pulses generate electromagnetic waves, and since those waves are brain waves, it could detect those brain waves.

Through rapid progress in brain science, BMI (Brain-Machine Interface) and robotics, people can become self-reliant insofar as their brains remain active. Such an era is bound to arrive soon. Concretely, this means that as long as a person's brain is active, brain science can help the person to eat or go to the toilet independently. Technology that provides dignity to individuals is bound to prosper.

Remarkable advances have also been made in medical care. Medical procedures to regenerate parts of the brain that have been damaged due to cerebral hemorrhage and other causes have inched closer to commercial use. Stem cells are injected and delivered to the affected area, and rehabilitation is performed by robots as shown in Fig 20. The procedure of regenerating the motor nerve system through the learning curve effect is already in the clinical trial stage. If it succeeds, medical care will likely undergo a transformation.

Participation of Active Seniors Is Indispensable

It is well known that elderly people's participation in society has a positive effect on them. Studies have suggested that this has had a positive impact on the reduction of their medical and nursing care expenses. More importantly, the participation of the elderly has become essential for society itself. At Mayekawa Mfg. Co., Ltd., employees who wish to continue working even after reaching retirement age can do so for as long as they wish, with the approval of people around them. In fact, the world's top-class refrigeration systems are developed at this company, and most of them are reportedly the product of joint work between postretirement age elderly employees and active-duty employees (Fig. 21).

This example has far-reaching implications. Active-duty employees not only lack experience but are pressed with their duties. They cannot afford to focus their efforts on the development of new products. Perhaps the empirical knowledge of the elderly and the leeway they have in their duties, or lack thereof, make up for the shortcomings of active-duty employees, thereby producing a synergistic effect.

Accordingly, there should be several opportunities for the elderly employees to utilize their skills more effectively. In the field of education, the lack of diversity in the experiences of the teachers has undermined their ability to educate students. If elders who have experience living abroad, have managed complaints, or have technical experience participate as teachers, the positive effects on the quality of education would be immeasurable. In this way, active seniors are important in helping us move toward a platinum society (Fig. 22).

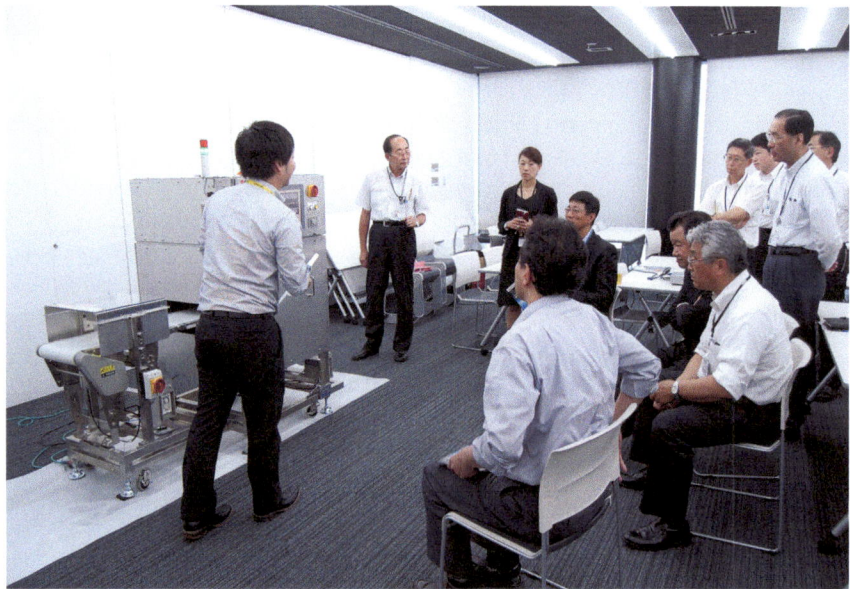

Fig. 21 Collaboration between active duty people and active seniors breeds creativity. (Courtesy: Mayekawa Manufacturing Co., Ltd.)

Fig. 22 Robot classroom taught by people who have experience working on control systems. (Courtesy: Next-Ph Robot Club)

Various Options

Companies moving their headquarters or select business functions from the city to regional areas have appeared. Komatsu Ltd. and YKK Corporation have transferred a part of their headquarters' functions to Komatsu City, Ishikawa Prefecture, and Kurobe City, Toyama Prefecture, respectively. They have reported that there has been no inconvenience at all since there is ready access to the Internet and the Shinkansen bullet train and airplanes are viable means of transportation. On the contrary, while the number of children of married women who work at Komatsu's headquarters in Tokyo is 0.9 on average, the figure is 1.9 in the Komatsu district. In other words, moving a company's business functions away from urban city centers can help serve as a countermeasure against a declining birthrate (Fig. 23).

Furthermore, Aratana is a venture company in Miyazaki City with 120 employees. The following comment by the company's young president is impressive: "Once you step inside the office, you will find a business environment that is no different from that in Tokyo (Fig. 24). Once you step out, you will find the enriching and beautiful living environment of Miyazaki." Through progress in information technology and means of transportation, local regions are no longer just the countryside. We have diverse options throughout the stages of our lives on where we want to work or live. In the future, there will be more options, and we will have greater freedoms. The problems of overcrowding in the cities and underpopulation in rural areas can be resolved simultaneously.

KOMATSU Way Global Institute -
(Above: Female employees
Right: Scene of female employees engaged in
childcare training)

Fig. 23 Komatsu Ltd. transferred education department from Tokyo head office to Komatsu City, Ishikawa Prefecture. (Courtesy: Komatsu Ltd.)

Fig. 24 Once you step inside the office, you will find a business environment that is no different from that in Tokyo. Once you step out, you will find the blessed living environment in Miyazaki. (Courtesy: Aratana)

Free Participation

Japan is known as a society where people have strong personal ties and networks. These interpersonal ties can be traced back to the fact that people were traditionally involved in collaborative work such as rice farming, and that Japan has always been able to develop by using its own resources, since it has never been invaded by other countries. When Japan entered the age of industrialization after the Meiji period, collaborative work at companies cultivated these ties. However, in recent times, there has been much talk going around about whether such ties are getting weaker.

I wonder whether ties between people can be maintained in the future. If serious collaborative work can foster ties among participants, the collaborative work involved in creating a platinum society should be able to assume that binding role.

Participation in collaborative work has been mostly compulsory in rice farming and at companies, and those who opposed solidarity have traditionally been thrown out of the pack. In other words, that was a participatory society with no freedom. On the contrary, a platinum society is a society in which all can freely participate.

For example, I wonder whether people will be commuting to work from Monday through Friday in 2030. That is most likely improbable. I think that on average, there will be a 3-day weekend, and people will be allowed to choose between commuting to work and working from home. If that is the case, raising children while continuing to work will become increasingly convenient, and likely even the concept of childcare leave disappears (Fig. 25). Companies that have adopted such flexible working systems have already started to appear.

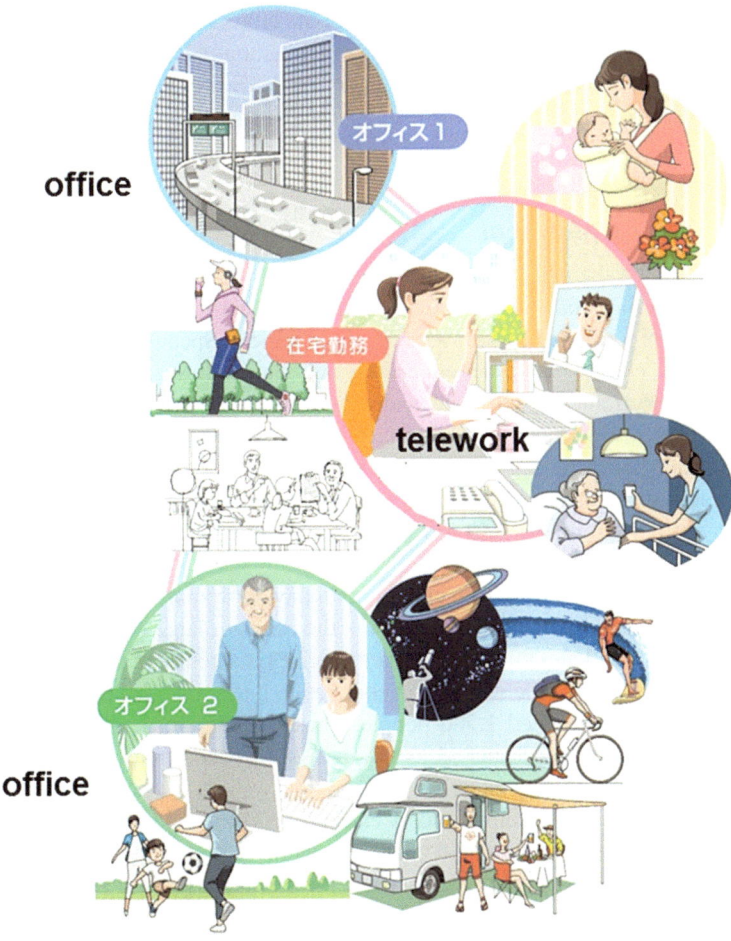

Fig. 25 Working styles have changed, with some people commuting to work two times a week and working at home for two days. (Courtesy: Platinum Society Network)

It will be a matter of personal choice for the elderly to continue to work in competition with active-duty employees or work like those at Mayekawa Manufacturing. Of course, they also have the option to enjoy retirement life in the countryside or in cities.

Education of Information Technology Nurtures the Next Generation

Information technology literacy has become a basic skill alongside reading and writing. In addition, good use of information technology will significantly increase the efficiency of education. For instance, videos are more convincing than a litany of words in explaining the miseries of war, art, or the mechanism of fertilization and birth. It is necessary to incorporate information into education for the sake of educating future generations, as well as to enhance the effect of education. Small groups of elite people with a strong will and young organizations such as the Association of Digital Textbook and Teaching (DiTT) and CANVAS are beginning to move the world in this sense.

Society with Employment

Employment is important not only to help individuals earn money but also functions as an individual's point of contact with others in society. Specifically, in the coming era when food, clothing, and shelter become daily necessities, as well as things such as consumer electronics, cars, music, and videos can be obtained at increasingly low prices, employment will be more important as a point of contact with society rather than to make a living.

If we stop to think about this, there is a concern that artificial intelligence may take away jobs. Will robots start to replace people's manual labor and even eliminate the need for intellectual labor?

Linkers Corporation, a venture capital company, has pointed us in the direction of the answer to that question. Trying to link the needs of large companies with the technologies of small and medium-sized enterprises is a common idea; they tried to create a system to get people involved in connecting one another's information. That number has already reached 1,700 people. These people are doubled linkers—in other words, they link needs and seeds through skillful coordination. Although this venture company created an excellent information system, the uptake was poor. In fact, failures of this nature committed by entrepreneurs are not unheard of. The idea can only work when there is enough uptake by people, suggesting that only collaboration between artificial intelligence and people can turn this into an efficient and comfortable system for humans.

There are thus hints to the answer in education, health, tourism, and services, all of which have become increasingly important in a platinum society.

The concept of "Society in the Loop" has been discussed in relation to artificial intelligence. It means that society participates in the loop of the development of artificial intelligence. If society participates and artificial intelligence evolves through trial and error, ethical issues as well as the fear that artificial intelligence may take away people's jobs can potentially be overcome.

Society Where Children Will Be Born

If asked to cite just one important issue confronting Japan, it would be the declining birth rate. Ethnic groups whose fertility rate falls below 2, or more precisely 2.05, will eventually disappear. Population projections are simple. That is because if 30 is the peak maternal age, those who will give birth within the next 30 years are women who are currently 0–30 years old, and that population size has already been determined. Taking that into consideration, the accuracy of the projection that the population will fall below 100 million to 97 million by 2050 is high. If we do not act now, Japan will clearly fall into a crisis because of the population decline and high senior ratio. Ultimately, what good is it to create a society in which young people do not want to have children? If people can freely choose how they want to work, how they want to raise children, how they want to live, have diverse employment opportunities, and no more anxieties about the future of our environment and resources, people will have hope for the future. Accordingly, they will probably hope to have a family and children.

A platinum society is a society where people who lead a normal life can have a family and raise children, if they so wish.

Knowledge Structuring Will Lead to Solutions

Innovation is indispensable in an era of transition, but it is not often the case that great inventions and discoveries to the extent of being worthy of receiving the Nobel Prize are necessary. Mankind already possesses a great deal of knowledge. For example, photosynthesis is a process in which light is used as the energy source to produce sugar and carbohydrates from CO_2 and water in the chloroplasts of plants. During the nineteenth century, this was the extent of our knowledge regarding photosynthesis. When we entered the twentieth century, however, greater intricacies have now been discovered in succession, down to precise molecular structures. As a result, we have now amassed a huge amount of knowledge that is possibly 10,000 times or even one million times more than before. Accumulation of such knowledge is progressing in all areas. Yet, while the accumulation of knowledge in itself is a good thing, no human being has a full grasp of the whole picture of this knowledge.

All too often, human beings do not know how to use the knowledge they have acquired.

If we can properly utilize appropriate knowledge from this vast knowledge base, we can resolve many issues. I call this knowledge structuring. From here on, if we can develop the methodology of knowledge structuring, take concrete actions, and link this with an entrepreneurial spirit, it will be possible to generate various innovations.

These innovations will occur in health, medical care, nursing care, harmony with nature, tourism, and education if we properly utilize the knowledge we currently possess, and through this, as a result we will be able to enjoy comfortable services that deliver high productivity.

How to Expand and Develop More Successful Examples

There are areas far from Tokyo such as Yanedan in Kagoshima Prefecture, Ama-cho in Shimane Prefecture, Kamikatsu-cho in Tokushima Prefecture, and Niseko-cho in Hokkaido that have created conditions that closely resemble a platinum society, despite setbacks of depopulation and stagnant growth. What is common among these places is the existence of exceptional leaders and excellent collaborators who skillfully leverage the region's unique characteristics. This however misleadingly gives off the impression that exceptional success stories under special circumstances that can hardly be copied.

However, even in an industrialized society, the horizontal expansion of factories has been no mean feat. The methodology for that purpose has been researched and summarized in the form of textbooks, handbooks, and manuals, and has supported the industrial age.

We must create a way to implement horizontal expansion of a platinum society. Using good examples as its source of reference, the Platinum Society Network (PSN) is trying to compile knowledge by summarizing the task of implementing horizontal expansion into a platinum concept handbook. Although it still cannot be said that this is sufficient, the PSN considers that it may possibly gain wide acceptance as an instructive methodology.

Challenging the Issue Through Knowledge Structuring and Action

There has emerged a group of scholars who boldly challenge difficult but important issues through knowledge structuring and action.

By gathering young researchers from eight universities to carry out a comprehensive project in Tanegashima, the Presidential Endowed Chair for "Platinum

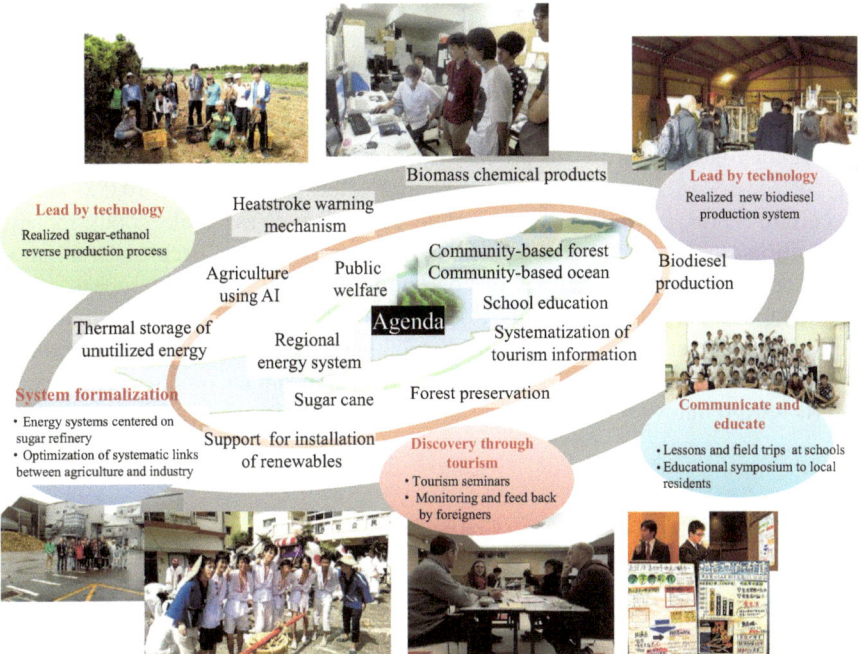

Fig. 26 Tanegashima Project – Seeking solutions to problems with knowledge structuring. (Source: Presidential Endowed Chair for "Platinum Society," The University of Tokyo)

Society" at the University of Tokyo has achieved great results. Fig. 26 shows an overview of this ambitious project.

What is important is to access and utilize the existing body of knowledge and integrate it toward resolving issues, through the means of knowledge structuring and building a relationship of trust between external human resources and local persons and organizations. Drinking together is not enough to build a relationship of trust and confidence. It is indispensable to present a comprehensive and concrete project that is convincing to all stakeholders, regardless of their experiences and interests. Based on the results of 3 years of efforts, great strides have been taken toward the construction of a general approach for realizing a platinum society in a given region.

With the activities of the Endowed Chair for "Platinum Society" serving as a catalyst, the Society of Chemical Engineers, Japan, has launched "The Working Group on Social Implement Engineering" and is trying to integrate knowledge for social implementation. The Engineering Academy of Japan has started collaborating with this movement not only the chemical sector, but also in the wider engineering sector to create a platform for cooperation in social implementation.

Where there is a need, we will create new businesses through knowledge structuring and develop them into a large industry, thereby allowing us to become more affluent. More than the increase in material possessions, it will be for realizing a better QOL, such as security and comfort.

Innovations from Mega-Cities

Although Yanedan and Ama-cho are successful examples in rural areas, mega-cities also play a substantial role in innovation. If innovation is to start from the intersection of heterogeneous ideas, its probability is proportional to the square of the density of the ideas. If the density is 100 times, the probability is 10,000 times. Accordingly, Tokyo is a place where various elements such as people, goods, money, and information are accumulated, and where the probability of occurrence of innovation is the highest. Indications of how innovations are generated appear in the Platinum Triangle connecting Shibuya, Futakotamagawa, and Jiyugaoka, situated in Tokyo's southwest district. It is in this district that a multiple number of companies in different industries as well as creators and people from universities are exerting efforts toward innovations necessary for sustainable new town development.

Viable Business Ensures Sustainability

Reviving Mishima City to be in harmony with nature as symbolized by fireflies populating the area is a typical example of how to aim for a platinum society. It has been significant in generating great value for its people both in terms of health and peace of mind, as well as by strengthening bonds and increasing the happiness of those involved. At the same time, the city's GDP has grown due to an increase in tourists. Rather than relying on subsidies, the activity stands on its own as a market economy. A platinum industry does not have to rely on a capitalist economy in which the return on invested capital must be maximized. However, to successfully sustain the path to a platinum society, many activities would have to be viable as businesses.

A Society of Lifelong Learning

Intense changes relating to work are inevitable during periods of transition. Unlike how it has been in the past, we can no longer expect to continue working in one job for our whole lives. As such, education for changing jobs would become a very important industry. In addition, a platinum society is one in which people learn from one another.

In a children's song about a sparrows' school, the school teacher trains students by singing, "Chi-papa, while swinging a whip as an orchestra conductor's baton, and another chi-papa, demanding still not good enough." This song describes universities in developing world. A platinum society is a future society that nobody knows about. As of yet, there are no teachers who have walked this path, no one who knows the exact route to a platinum society. To illustrate this point, another chil-

Fig. 27 Businesspersons in the Marunouchi district neighborhood are studying voluntarily aside from work. (Courtesy: Marunouchi Platinum University)

dren's song about a killifish school has the following lyrics: "Regardless of who are the students and who is the teacher, everybody is playing together." This indicates that we are entering an era similar to the killifish schools. Where there are no teachers, we should rely on exchanges of the diversity of experiences and ideas of the people who are participating. In so doing, we will transition into a society where people will continue to teach one another, train one another, and learn from one another.

The Marunouchi Platinum University is like a killifish school as it enhances the ability to move toward a platinum society (Fig. 27). Although it only recently opened in July 2016, it is already packed with students. It is encouraging to know that we have entered an era where businesspersons in the Marunouchi district neighborhood subscribe to these paid courses.

The Platinum Society Network consistently positions human resources development as one of its most important projects (Fig. 28). The Platinum Society School is diverse. It is another killifish school aimed at building a society focused on lifelong learning. The many examples of their initiatives include a university where local government employees gather in Tokyo to present "Our Town's Platinum Concept," a policy competition event for employees' groups organized by local governments, a venue that gathers public health nurses who play a key role in regional medical services, and a Platinum Energy School that promotes citizens' understanding through junior and senior high school students. Even the "Platinum Development of Future Human Assets Juku (Cram School)," where Nobel Prize scholars and former Bank of Japan governors teach junior high school students, as well as robot

Fig. 28 Human resources development is the most important business. (Courtesy: Platinum Society Network)

classes where active seniors teach elementary school children. These are like the killifish schools in which even the teachers themselves find themselves learning. It would not be surprising if the education industry will be hiring the largest number of employees in 2050.

Developed Countries Can Achieve Economic Growth

In the past, only a small portion of the privileged class could afford to ride on horse-drawn carriages or palanquins, while public traveled on foot. Even when automobiles appeared, the situation remained unchanged. However, when Henry Ford developed a standardized mass production system with the goal of "making automobiles affordable for all employees," and as a result productivity dramatically improved, prices of automobiles were reduced to a tenth, resulting in 100 times more people were buying cars. This caused a positive spiral that expanded the market size by tenfold. Such is the basic relationship between innovation and economic growth. However, does that mean that there are no more seeds of economic growth in developed countries, in which people already have a given level of material wealth?

That is not the case. It could make up some of the necessary conditions of a platinum society shown in Fig. 9.

Civil engineering and construction have so far been considered a typical 3D (difficult, dirty, and dangerous) industry. Currently, cutting-edge companies are focused on i-Construction, where everything from design to maintenance is informatized, drones perform surveying, and robots and automatic tractors complete the work. If successful, the unit price of the construction work would be reduced to a tenth. This means it would be possible to perform 100 times as much construction work, though such work will probably not exist in developed countries. In reality, there exist enormous needs in this area, particularly in the area of maintenance. Infrastructure such as water and gas, bridges, tunnels, and expressways have aged. It is said that it would require 100 trillion yen a year for the maintenance of all the above-mentioned infrastructure, and virtually nothing has been done. If this continues, the future is bleak. However, a better future is possible if productivity increases tenfold. What we need are innovators like Henry Ford.

Furthermore, if one caretaker can look after ten times as many people, nursing care costs would be reduced to one-tenth many could receive nursing care services without anxiety, causing the market size to expand. The same would hold true for medical care. Given that expensive cutting-edge medical care is considered problematic in that it may lead to financial collapse, what needs to be done is to reduce costs and improve productivity. As Japan moves toward an aging society, medical care needs will swell to more than 10 times its current rate. By training co-medical human resources in large numbers and introducing information systems including artificial intelligence in earnest, it will be possible to improve productivity. All we need is innovation.

GDP and IWI

Henry Ford responded to the needs of the people by enhancing the productivity of automobiles, thereby spurring economic growth. However, even if creative demand in a platinum society generates innovation, it may not necessarily lead to economic growth through business, due to the problem of externalities in the economy.

For example, if people worked together in groups to provide nursing care to their parents, that would not be counted in the GDP, even if it were high-quality care and the productivity is not inferior to that of the nursing care industry. This is similar to the case when eating at a restaurant would be counted in the GDP, but if people invited each other to dinner at their homes, that would not be counted in the GDP. In addition, health and self-reliance are necessary conditions of a platinum society. In the current market, however, prevention cannot be counted in the GDP. People suffering from diabetes increase the GDP if they become dialytic, but preventing them from becoming dialytic would instead reduce the GDP. Dialytic patients are forced to suffer a drastically lower QOL, but they significantly increase the GDP. This is similar to the pollution problem. A mechanism to internalize these externalities into the economy should be built as soon as possible (Figs. 27 and 28).

Many of the activities directed toward a platinum society consist of providing services to enhance QOL. Manufacturing and selling automobiles and television sets outside of the market economy are unlikely, but it is not unlikely for services to be performed outside the market economy. However, it should be possible to internalize those activities and turn them in the direction of increasing the GDP.

In the era in which society developed and the economy grew in developing countries because of their material demands, GDP was an appropriate social indicator. However, the index to enhance the QOL must of course be that which represents quality. The United Nations has proposed the IWI (Inclusive Wealth Index), which is a comprehensive indicator that includes the environment and sustainability.

In a platinum society, we should think about the indicators of economic scale and QOL—for example, both the GDP and IWI—as important indicators.

Platinum Industry and Economic Growth

In Fig 29, each dot corresponds to one field of industry, with the fields arranged from left to right in order of CO_2 emissions. The vertical axis represents the amount of CO_2 emissions per ¥1 trillion worth of added value for a given field of industry, whereas on the horizontal axis, the distance from the near point to the left represents

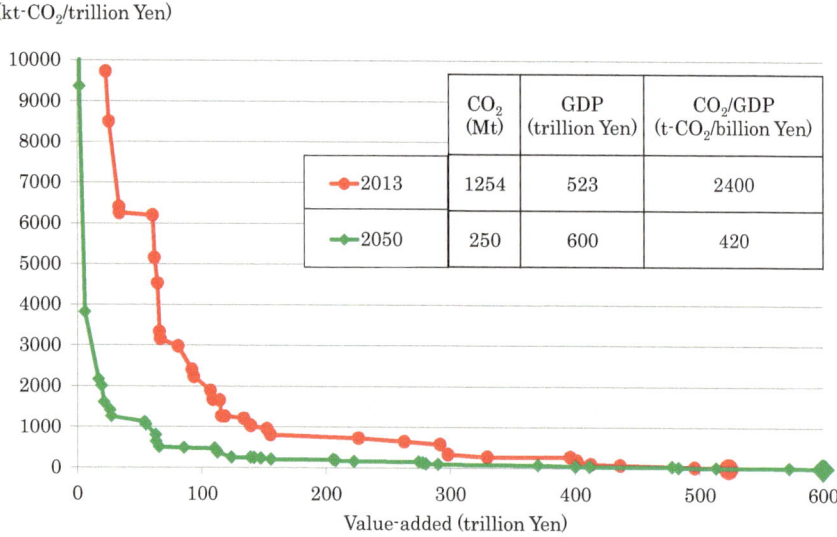

Fig. 29 Economic structure of bright low-carbon society. (Source: Created based on materials from the Center for Low-Carbon Society Strategy, Japan Science and Technology Agency)

the amount of added value of that field of industry, or in other words, its contribution to the GDP. The GDP in 2013 was ¥523 trillion, as represented by the dot at the rightmost end, while the amount of CO_2 emissions was 1.254 billion tons shown by the area below the line. Industries such as steel and chemicals appear at the leftmost end of the horizontal axis, indicating that they generate large amounts of CO_2 per GDP. Positioned at the right are tertiary industries such as financial, real estate, commercial, medical care, education, and services. This specifically indicates that the tertiary industries emit less CO_2 compared to the secondary industries.

The only way to achieve low-carbonization and simultaneously increase GDP is to achieve low carbonization of the industries at the upper left and enable low-carbon-emitting industries at the lower right to grow. In other words, low carbonization of the industrial and transportation sectors must be achieved, while promoting growth of such areas as education, tourism, and medical care. A platinum society promotes the use of electric furnaces in the steel industry, zero carbonization of automobiles, and low carbonization of industries on the left side such as renewable energy, as well as growth of areas such as for health and self-reliance, education, and tourism. In other words, the transition to a platinum society is also a means to achieve low carbonization and economic growth.

For example, if we were to aim at an annual growth of 0.3% until 2050 and reduce CO_2 emissions by 80% by this point, the GDP in 2050 would need to increase to ¥600 trillion and CO_2 emissions must be kept at 251 million tons. One case in which this target value can be achieved is represented by the line at the bottom. As outlined in this chapter, while it cannot be accomplished easily, it is nonetheless possible. Furthermore, as it has happened with renewable energy and automobiles in the past 20 years, innovation can be expected to occur if the goal is in reach.

Adapting the industrial structure to a platinum society and achieving both quantitative growth and dramatic enhancement of quality—that is the right choice.

Developing Countries Aim at a Platinum Society Together

Developing nations have the advantage of being able to begin from ground zero. Africa did not have hardwired phones, allowing for a world of mobile phones and internet to come to pass immediately. In building a system of renewable energy, developing nations have an advantage of not having existing electric systems—i.e., an electric grid that presupposes the existence of large-scale power plants, which could be restrictive. There are also advantages of urban mines. Developed countries already have established steel industries that operate on shaft furnaces, causing difficulties when transitioning to utilizing metal scrap materials instead. While developing countries do not produce scrap materials since it has had a short time to build up man-made objects, it is nevertheless able to build the technique and system centered on scraps by importing them from developed countries because they do not have well-established shaft furnace-centered industries. This system is that of a platinum society and can be exported to developed countries. In other words, developed and developing countries can collaboratively lead mankind toward urban mining. Because this is transition period, there are huge advantages of starting from scratch.

The second advantage is that much infrastructure and products that must be imported are now reasonably priced. Renewable energy has already become the cheapest source of energy, insofar as we make the appropriate choices for a given locality. Energy-efficient home electronics are also no longer uncommon, which was not so just 5 years ago. Today's pace of change is advantageous for developing countries.

The third advantage is the huge capital that is flowing throughout the world just seeking causes to invest in. Many of the goods a developing country must import rely on initial investments. Systems such as energy-economizing heat pumps, renewable energy such as wind power and solar cells, and health assistive systems that make use of an assistive robot mostly require an initial capital investment, and only require a recovery of funds later. This seems disadvantageous for developing countries with little accumulated capital. In fact, short-term capital does not favor such fields. However, there is massive capital in the world seeking low-risk stable returns such as pension funds. If we can direct such capital toward appropriate investments, the difficulty for developing countries to acquire funding would ease significantly.

The fourth point to note is pollution issues, as exemplified by China's unsustainable levels of PM 2.5. This is an example that a developing country should absolutely avoid, especially because it was completely predictable, with mankind knowing the cause and solution all along. Chinese academics were aware of this too. If they had dropped their rate of economic growth to 9% and allocated 1% to investing in the environment when their rate of economic growth was 10%, this situation would have been avoidable. We should let this be a lesson for us. What we need is a way to incorporate knowledge into policy.

Summarizing the above four points, developing countries should not blindly follow the path of developed countries. If they do, they will repeat the mistakes of developed countries, and suffer pollution and lifestyle-related diseases. They should directly focus on the Platinum Society. Developing countries have advantages that developed countries do not have, and those advantages are complementary. Developing countries will realize the best things, which are then spread to the world including to the developed countries. Developing countries and developed countries must think of one another as partners in inching the world closer to realizing a Platinum Society.

Mankind has declared through the SDGs that "no one will be left behind," and have achieved consensus on the Paris Agreement. Now, all we need to do is act. It is understandable that developing countries want to chart their own paths independently. Developed countries should collaborate among themselves for a platinum society together. Developing countries should respond by developing and streamlining social conventions, social systems, and securing law and order, so that they can collaborate with developed countries.

The key to accomplishing this is human capital, which is mainly achieved through education. Africa's population challenges are one of the foremost challenges in the world's population explosion, and the answer lies in education. How Africa ends up will determine the future of mankind, not least the population issue. The spirit of "the good you do for others is good you do yourself" is the hardline attitude that must now grip all nations.

The solution is becoming clearer concerning issues facing both developed and developing countries. What is being asked of us is our courage as humans to collaborate and work toward that solution.

Social Disparity and Social Security, the State and the Market

The Earth has the capacity to provide "things" to allow all people, including those born during the future population boom, to live a comfortable life. We can all achieve a sustainable life that is materially abundant and satisfying at no one's expense, while working within Earth's limitations. In other words, the Platinum Society is a vision that can be realized.

However, to arrive at this point, there are issues that we must come to terms with. In society, fairness and equality are significant values, and yet they face worldwide challenges, as exemplified in Japan through the large disparity between full time and temporary workers. The social security system has eased the large social gap in society. In reality, however, we are unable to fund this solely with our insurance income, and our government debt has only grown. Since there is a limit to how much the working generation can shoulder insurance payments, we can only reduce debt by increasing tax revenue. Therefore, we must implement a total reform of taxes and social security.

We need to increase the earnings of corporations and individuals to increase tax income. For that, we need innovation. Industrial policy has traditionally been used to support specific industries such as cars and solar power generation, but this is no longer relevant for us. If we combine the technology and knowledge we have today, we can develop most things. For instance, to improve the emergency transportation system, we used to concentrate on developing a specialized ambulatory vehicle and system, but going forward, we can build an efficient system through combining existing techniques adapting smart cars and drones, and collaborating with dynamic maps. To realize such innovations, the government must think about how to build a lateral network among various industries and technologies. Only the government can initiate such a transformation, although the government's vertical bureaucratic structure may stand as a formidable impediment to its implementation in many industries.

A System for Innovations

Dr. Hiroshi Yoshikawa, who is featured in an interview at the end of this book, argues in his latest book *Demography and the Japanese Economy* that economic growth has been brought about by innovation rather than by an increase in the size of the workforce. Seeds of innovation can be cultivated by the Platinum Society and its buds are already bearing fruit in many places.

Innovation stems not only from the making of goods but also by making them more affordable. The basic mechanism of a free market is one like Henry Ford's, where he shaved the price of cars to a tenth, allowing for 100 times the number of people to buy cars, thereby expanding the market by tenfold. However, for instance, even if advanced medical and assistive robots are invented, companies will have little motivation to lower the price if the current social security system must fund them. This would cause considerable financial strain on the government. That would be akin to saying that expensive cars should be distributed to citizens because they are great. The necessary condition for innovation is a system in which the companies are motivated to lower prices. To realize the SDGs aim of leaving no one behind, it would be effective to work through the free market.

While the capitalistic system based on democracy appeared to have triumphed after the end of the Cold War in the twentieth century, this system is now facing challenges. We must find answers within the finite limitations of Earth's resources while wisely utilizing the advantages of a free market.

Twenty-First Century Is a Turning Point

From avoiding population explosion to overcoming a declining birthrate, abolishing hunger to preventing obesity, fossil resources to renewable energy, mining resources to urban mining, overcoming pollution to harmonizing with nature, high-speed production to leisurely living, compulsory participation to free participation—the twenty-first century marks a great turning point in human history. The issues that we face today may be resolved by fully utilizing the freedom we have achieved through civilization. We will achieve resource self-sufficiency, harmony with nature, health and self-reliance, lifelong learning, freedom of participation, a rich culture, and diverse options, when we resolve these issues. In other words, we will be working toward a platinum society.

Platinum Society Network

The vision for the twenty-first century is a platinum society. Its buds are already sprouting in many places. However, what should we do to work toward a platinum society from the society that we currently live in? We have little time left. What kind

プラチナ構想ネットワーク
第1回シンポジウム

Fig. 30 Platinum society network

of efficient methods can we find? What we do know for sure is that it will not begin until people start acting on their own accord. What we are aiming at is a platinum society that is an autonomous, distributed, and collaborative system. With that objective in mind, we have no choice but to experiment, even if we experience setbacks and leaps, and be determined to make progress among independent groups, by exchanging information, presenting ideas, and structuring knowledge to create a methodology that leads to success. At the very least, we need a group of people who have decided to head in that direction. It is not clear specifically what the Platinum Society Network can do, but for a start, people who are single-mindedly focused on heading a platinum society have already come together to establish the Platinum Society Network in August 2010 (Fig. 30). Its philosophy was established in the initial 3 years, and specific activities were carried out in the following 3 years. During this period, the initial membership of 46 groups expanded to 300. Going forward, we intend to expand our membership pool. Of course, we cannot do this alone. Fortunately, we have been able to strengthen collaboration with groups that are aligned with our vision. In other words, ours is a "network of networks" that boasts an action plan which applies to autonomous, distributed, and collaborative systems.

My favorite words: "the best way to predict the future, is to actualize it." Let's march toward a Platinum Society together.

Tokyo, Japan   Hiroshi Komiyama
                                                               Koichi Yamada

References

Arrow KJ (1951) An extension of the basic theorems of classical welfare economics. In: Neyman J (ed) Proceedings of the second Berkeley symposium on mathematical statistics and probability. University of California Press, Berkeley, pp 507–532. https://projecteuclid.org/download/pdf_1/euclid.bsmsp/1200500251. Accessed Jan 2018

Erin Magee, USAID/OFDA (2015) https://www.usaid.gov/crisis/micronesia. Public domain via Wikimedia Commons. Accessed Jan 2018

Irish Defense Forces (2015) https://commons.wikimedia.org/wiki/File:LE_Eithne_Operation_Triton.jpg. Licensed under https://creativecommons.org/licenses/by/2.0/. Accessed Jan 2018

Kates RW (1999) with National Research Council, Board on Sustainable Development, 1999, Our Common Journey: A Transition Toward Sustainability. National Academy Press

Kates, RW (2011) What kind of a science is sustainability science? Proc Natl Acad Sci 108(49, 99):19449–19450. http://www.pnas.org/content/108/49/19449.full. Accessed Jan 2018

Kauffman J (2009) Advancing sustainability science: report on the international conference on sustainability science (ICSS) 2009. Sustain Sci 4:233–242

Komiyama H (1990) Chikyuu ondanka mondai hand book. IPC

Komiyama H (1995) Chikyuu ondanka mondai ni kotaeru. UP Sensho

Komiyama H (1999) Chikyu Jizoku no Gijyutsu. Iwanami Shinsho, Japan

Komiyama H (2001) Chikyu Jizoku no Gijyutsu (Chinese Version). China Environmental Science Press, China

Komiyama H (2011) Nihon Saisouzou. Toyo Keizai Inc Komiyama H, Matsushima K. (2012) Platinum Vision Handbook —Change the world through the power of active elderly. Platinum Vision Committee, Japan

Komiyama H, Kraines S (2008) Vision 2050—roadmap for a sustainable earth (originally published in Japanese, 1999). Springer, Japan

Komiyama H, Takeuchi K (2006) Sustainability science: building a new discipline. Sustain Sci 1:1–6

Meadows DH et al (1972) The limits to growth, 1st edn. Universe Books, New York

Michael Schwarzenberger (2014) Solar Panels. Licensed under Public Domain C0

Porter ME (1991) America's Green Strategy. Sci Am 264(4):168

Winchell Joshua, U.S. Fish and Wildlife Service (2013) A close shot of wind turbines wind farm. Public domain via Wikimedia Commons

World Commission on Environment and Development (WCED) (1987) Our common future. Oxford University Press, Oxford

Contents

Chapter 1
The Message in "Vision 2050"

1.1 Behind the Birth of Vision 2050

1.1.1 The Need for a "Macro" Vision

In 1999, as the world grew excited over the start of a new millennium, I published a volume titled *Chikyū jizoku no gijutsu* (Iwanami Shinsho) – "The technology for global sustainability." In the second half of that decade, the signing of the Kyoto Protocol had helped to make people everywhere more broadly familiar with the problem of global warming, and momentum began building on reducing CO_2 emissions. It was also around this time that many began to sense that the material civilization that had supported the dramatic growth of the twentieth century had reached an impasse. People conscious of these matters began taking action in an effort to avoid the difficult state of affairs that the exhaustion of energy resources, global warming, the generation of large amounts of waste, and similar developments present. However, the data and information that provide the grounds for action were lacking, and discussion likewise had not yet reached a state of maturity.

For people to take action requires them being able to believe that the actions they are attempting are correct and are being done for the world's sake. However, attempts to encourage recycling have been met with people expressing the view that it merely adds to costs and is unrealistic. Meanwhile, studies on the adoption of solar batteries have led some specialists to reject them, saying that the payoffs are not proportionate to the costs and the absolute amounts of energy achieved are insufficient.

Finding a middle ground among views that directly contradict one another like this will require a "macro" vision. Such a vision would be one that everyone can share, that has a consistency as a whole that fully takes into consideration various crucial, individual items. If there is a shared sense of what the future should look like, then it will establish the orientation that people should take moving forward as a matter of course regardless of conflicts over separate arguments like those over

© The Author(s) 2018
H. Komiyama, K. Yamada, *New Vision 2050*, Science for Sustainable Societies,
https://doi.org/10.1007/978-4-431-56623-6_1

recycling and solar batteries. Regardless of the varied and diverse issues that are brewing, I believe that there was no shared sense of what shape the future should take.

1.1.2   An Affluent Lifestyle for All

It was with this sense of the problems facing us that I formulated my "Vision 2050" and put it forth in *Chikyū jizoku no gijutsu*. Vision 2050 represents an image to be achieved by the mid-twenty-first century of a material circulating society that relies on renewables in an energy-efficient way. It is based on the three principles of "saturation of man-made objects," "improved energy efficiency," and "developing renewable energy." My premises when I formulated this were to see everyone in the world achieve a standard of living on par with what developed countries enjoy today, while at the same time solve environment- and resource-related problems. In addition, I used the most leading-edge data I could obtain at the time of writing, while taking up such separate topics as making vast improvements in energy efficiency and expanding the use of renewables.

To say "everyone on the planet will enjoy material abundance at the level of developed countries today" was an idea that nearly everyone in the world at the time held to be idealized but an unrealistic theory, and that the planet would not be able to bear if attained. Even today, most people still probably think so. However, did not the United Nations pledge that "no one will be left behind" through its sustainable development goals in 2015? Is there some vision of the world conceivable other than this? For the U.N. to issue such a pledge at the moment when I was thinking about Vision 2050 was unexpected, and it convinced me that I was not wrong. With that in mind, when I logically analyzed the results of all the research I had undertaken about lifestyles and society as of 1995, trends in population and man-made objects, technologies and their future, and the like, I realized that this was not an idealized theory by any means but a feasible vision.

1.1.3   Why a Low-Carbon Society?

The global environment is changing at breakneck speed. According to the Fifth Assessment Report (AR5) issued by the Intergovernmental Panel on Climate Change (IPCC), the average annual temperature of the earth's surface (both land and the ocean's surface) has risen 0.85 °C between 1880 and 2012. Furthermore, the change projected for the end of the current century based on data for 1986–2005 is for the average temperature to have risen by 0.3 °C at minimum and by as much as 4.8 °C.

Population growth is one factor behind the rising temperature. Over the 100 years of the twentieth century, the world's population rose from 1.7 billion to 6 billion.

Furthermore, owing to the modernization of people's ways of life, the quantity of all the material resources they needed grew and so the volume of agricultural output increased 7.5 times and that of industrial output by 20 times.

In concert with this, the carbon dioxide (CO_2) concentration in the atmosphere rose. The average concentration of CO_2 in the world a century ago was around 290 ppm; in contrast, observations made at the end of 2015 by the greenhouse gases observing satellite Ibuki showed the concentration to have exceeded 400 ppm.

The atmosphere serves to keep the temperature of the entire globe as a whole at about 15 °C, but when the CO_2 concentration rises that equilibrium is upset. The temperature difference may appear to be rather small when you look at the numbers. However, the planet is definitely warming, and that will have incalculable effects on the environment. When temperatures rise, ocean currents and airstreams change and so does the climate. Europe has experienced record-breaking cold weather, while on the other hand, Australia's dairy industry has taken a major hit from droughts. Japan, too, has seen a rise in both extremely hot days and heavy snowfall-induced damage. These developments are believed to be signs of climate change caused by global warming. If temperatures rise, the resulting situation will not be one that can be dealt with simply by dressing for summer year-round.

The world's population is forecast to rise by more than 2 billion from today's figures to reach 9 billion by 2050. Currently, the annual volume of CO_2 emissions per person in developed countries stands at about 8.4 tons. If all of humanity is leading lives on par with that of such countries by 2050, the volume of CO_2 emissions will reach 75 billion tons per year. By my research team's provisional calculations, the concentration of CO_2 in the atmosphere will be 600 ppm. Even compared to the change in concentration over the past 100 years, that which will occur in the next 40 is overly rapid.

The IPCC's worst-case scenario of a maximum rise of 4.8 °C assumes that no additional measures are taken to combat warming beyond those of today, while the best-case scenario of 0.3 °C presumes that very strict measures are taken. It goes without saying which scenario is better. In order to maintain the global environment, humanity should be aiming for achieving a low-carbon society that reduces the volume of CO_2 emissions as much as possible.

1.1.4 The Threat of Global Warming

Each of the reports the IPCC has issued have included an assessment of the effects that human activities have on warming. The First Assessment Report (FAR) issued in 1990 limited itself to mild comments to the effect that there were concerns that anthropogenic greenhouse gases could cause climate change. However, the expressions used have grown stronger with each report, with AR5 saying in essence that it was extremely likely that the primary factor behind the warming in the second half of the twentieth century was due to human influences. The degree of certainty is greater than 95%.

Temperatures have already risen by 0.85 °C and are projected to rise anywhere from 0.3 °C to 4.8 °C in the next half century. Humanity has never experienced such a dramatic change.

The risk due to warming that people are most familiar with is a rise in sea levels produced by the melting of the polar ice caps. It will take time for such an enormous amount of ice to melt, but once the ice is reduced and ocean temperatures start to rise, it will make it even easier for further melting of the ice. Even if the rise in atmospheric temperature is halted, it will require an enormous amount of time for the ocean temperatures to continue to cool enough for ice to form once again.

It is said that if the sea levels rise, then various islands such as Tuvalu in the South Pacific will sink into the sea. Research indicates the possibility that countries like Thailand with coastal areas may, depending on conditions, see considerable parts of those regions sink out of sight. Japan, too, will not go unaffected. If sea levels rise by around 60 centimeters, it is expected the belts of coastal regions around Tokyo, Osaka, and elsewhere that currently stand at exactly sea level would increase by 1.5 times. This would require social capital investments on measures to cope with flooding and water exposure.

Global warming will also bring about climate change. Even though Europe sits at a higher latitude than Hokkaido, it has a temperate climate. This is due to the general circulation of ocean currents wherein the warm Gulf Stream flows north across the Atlantic Ocean to reach Europe and the waters subside there. However, as warming proceeds, the amount of precipitation will increase and the saline concentration in sea waters will fall. If the specific gravity is not high enough, the waters will not subside to the ocean floor and the flow of the Gulf Stream is quite capable of stopping.

Should that happen, it is possible that Europe will become quite cold even if the Earth's temperature on the whole rises. Europe would experience a wave of extremely cold weather that would set new records for people dying in the thousands due to electric power and other essential utilities being cut to shreds. The social infrastructure takes the meteorological and climate conditions as they have been to date as its premise, so countermeasures will not keep up in the event of a drastic change.

Furthermore, if the Gulf Stream stops flowing to Europe, its heat will remain in tropical regions and could cause Mexico and the southern U.S. to heat up intensely. Hurricane Katrina, which practically wiped out the city of New Orleans, is thought to have been produced by this sort of mechanism.

Another terrifying projection is that of positive feedback in warming. Lurking beneath the surface of the Siberian permafrost is methane hydrate, with a structure comprising methane molecules that are surrounded by water. Warming will cause the ice to melt, releasing methane gas into the atmosphere. Methane has not received as much attention since the volume of CO_2 emissions has been much higher, but it has a greenhouse effect that is 10 times that of CO_2 per unit volume. This release of methane into the atmosphere through warming could generate feedback in that the greenhouse effects of that methane would cause warming to proceed all the more.

We know from recent climatological research that the global environment exists in a truly delicate balance. Even one slight change to one component in the atmosphere could change the ecosystem dramatically.

1.2 What Is Vision 2050?

1.2.1 The Vision for 2050

To keep global warming in check will require keeping down the CO_2 concentrations in the atmosphere. In Vision 2050, I ran simulations for the supplies of materials and energy as well as CO_2 concentrations for 2050 based on the volume of energy used in 1990.

In 1990, humans consumed a total amount of energy equivalent to 7.5 billion tons of resources, comprising roughly 6 billion tons of fossil resources and 1.5 billion tons of non-fossil resources (biomass, hydraulic power, and nuclear power). The world's population was around 6 billion people at that time, meaning each person consumed about one ton of fossil resources. There is in fact a considerable difference between developed and developing countries, with developed countries having consumed about 4.5 billion tons. Breaking down the amount of consumption per person by country, we see the figures were 2.4 tons for Japan, 2.5 tons for the UK, 2.6 tons for Germany, and 5.3 tons for the United States. If the figure for the U.S. is excluded as an outlier, then the average amount consumed per person in the developed countries of Europe and Japan was 2.3 tons.

The total population of the world is forecast to be 9.3 billion people in 2050. Of this, the population of developing countries will account for about 8 billion people. If they lead lives on par with those of developed countries, then multiplying that 2.3-ton figure by 8 billion people results in approximately 18.5 billion tons of fossil resources being consumed. Biomass will amount for 1 billion tons of this just as in the benchmark year. Also, if demand in developed countries in 2050 remains the same as in the benchmark year, then this means 5 billion tons of energy – comprising 4.5 billion tons of fossil and 0.5 billion of non-fossil resources – will be consumed. The global total for energy consumption will be 23.5 billion tons, equivalent to triple the figure for the benchmark year. This is what's behind the tacit understanding of many that the planet will not bear up if all people on it become affluent.

However, these calculations are premised by the technology of the benchmark year. If the technology evolves, then energy use efficiency will improve. For example, if energy use efficiency were tripled by 2050, then the amount of energy consumed would end up around the same as the benchmark year at 7.5 billion tons (Figs. 1.1 and 1.2).

Even so, if the amount of fossil resources used is the same as it was in 1990, then CO_2 would also continue to be emitted at the same levels. It thus would not be possible to return the global environment to conditions prior to that time. To arrest the

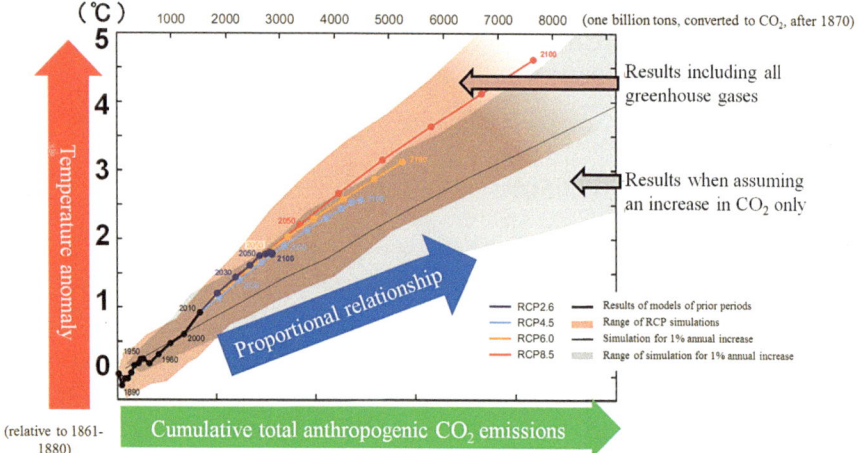

Fig. 1.1 Amount of rise in global average temperature as a function of the total cumulative global CO_2 emissions. (Source: Created based on Fig. 10 in the SPM of the IPCC Fifth Assessment Report)

progress of global warming will require even greater efforts than just improving efficiency.

In Vision 2050, I proposed increasing the ratio of renewable energy. The amount of energy use based on non-fossil resources in the benchmark year converted to carbon stood at approximately 1.5 billion tons. If this figure could be doubled, then the amount of fossil resources used in 2050 would end up at 4.5 billion tons. That is to say, if we could triple energy use efficiency and double our use of renewable energy we could keep the amount of fossil resources used down to three-quarters of the amount in the benchmark year. If Vision 2050 could be achieved and further progress is made in the latter half of the twenty-first century on improving energy use efficiency and expanding the use of renewables, then it might be possible to achieve a scenario in which the amount of fossil resources used is kept even further down.

What effect would that have on the CO_2 concentrations in the atmosphere?

In the benchmark year, the concentrations stood at 369 ppm. We have seen the atmospheric CO_2 concentrations rise in proportion to the volume of anthropogenic CO_2 emissions. The amount was increasing by 2 ppm annually as of the late 1990s. If it continues at that pace, the concentration will have risen 100 ppm in 50 years to exceed 469 ppm in 2050.

Meanwhile, if as noted earlier developing countries have attained a living standard on par with developed countries by 2050, and there is no change to either energy use efficiency or the proportion of renewables used, then the amount of fossil resources consumed annually will reach 22 billion tons. Calculations show that the CO_2 concentrations would reach 600 ppm at this point (see Fig. 1.3b). That figure would be far more than double the CO_2 concentration present before the Industrial Revolution.

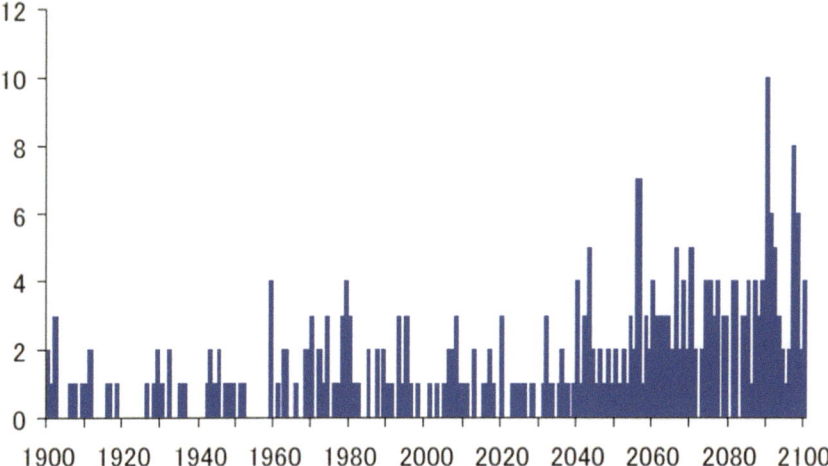

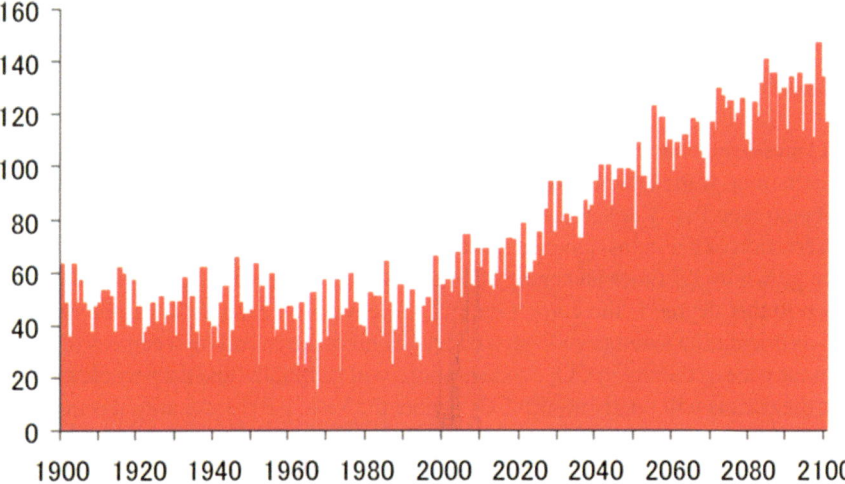

Fig. 1.2 Figure above; Change in the number of summer days with torrential rains in Japan. Change in the number of summer days (June–August) with torrential rains calculated in Japan between 1900 and 2100 (results for 2001 and later use the "A1B" scenario). If even one the cells covering the Japanese archipelago (approximately 100 km × 100 km) has a daily precipitation exceeding 100 mm, it is counted as a day with torrential rainfall. Since it is based on average values over broad areas, absolute values cannot be directly compared with observation data. Only relative changes are significant. Figure below; Change in the number of hot days (maximum temperature > 30 °C) in Japan (figure below). Change in the number of tropical face calculated in Japan between 1900 and 2100 (results for 2001 and later use the "A1B" scenario). If even one the cells covering the Japanese archipelago (approximately 100 km × 100 km) has a maximum temperature exceeding 30 °C, it is counted as a tropical day. Since urbanization has not been taken into account, and since it is based on average values over broad areas, absolute values cannot be directly compared with observation data. Only relative changes are significant. (Source: Center for Climate System Research, The University of Tokyo; the National Institute for Environmental Studies; and JAMSTEC Frontier Research Institute for Global Change)

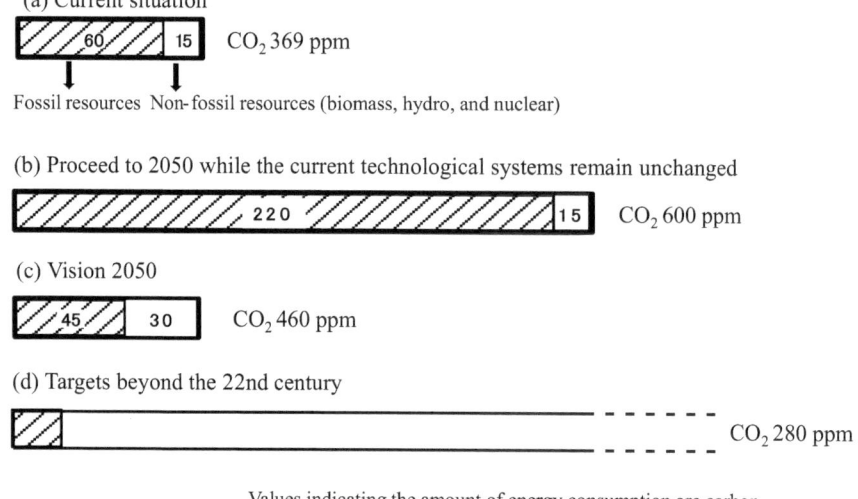

(a) Current situation

60   15 CO_2 369 ppm

Fossil resources Non-fossil resources (biomass, hydro, and nuclear)

(b) Proceed to 2050 while the current technological systems remain unchanged

220   15 CO_2 600 ppm

(c) Vision 2050

45   30 CO_2 460 ppm

(d) Targets beyond the 22nd century

CO_2 280 ppm

Values indicating the amount of energy consumption are carbon
equivalents (100 million tons)

Fig. 1.3 Energy scenarios and CO_2 concentrations. (Source: "Technology for Global Sustainability")

Conversely, under Vision 2050 with the amount of fossil resources used at three quarters that of the benchmark year, the CO_2 concentration is calculated to stand at 460 ppm (see Fig. 1.3c). While that figure is not all that different compared to the preservation of the status quo model (469 mm), if we consider that the population will have grown by 1.5 times and energy demand will have tripled, then keeping the level down to this degree is the correct option. Moreover, if Vision 2050 can be realized and its orientation maintained from the second half of the twenty-first century onward, then the pace of the rise in CO_2 concentrations will be blunted and drop to a level such that the oceans will finally absorb CO_2. Under the Vision 2050 scenario, it is even not impossible that the CO_2 concentrations could be returned in the twenty-second century to the 280 ppm level of the pre-Industrial Revolution world (see Fig. 1.3d).

The three objectives of Vision 2050 may be summed up into the following three items.

1. Create a material-circulating system
2. Triple energy use efficiency
3. Double the amount of renewable energy

1.2.2   A Happy Low-Carbon Society Is Achievable

Vision 2050 is meant to reduce the CO_2 concentrations in the atmosphere, but it does not impose excessive compromises or sacrifices toward that goal. It seeks to realize the three above objectives not through idealism but rather with scientific technology.

1.2.3 Saturation of Man-Made Objects and the Material-Circulating System

First, saturation of man-made objects is the key point of a material-circulating system. Man-made objects such as steel and cement continue to accumulate. They will fill every nook and cranny relative to the population and ultimately reach saturation. Demand is not something that will grow forever, even in China or India. Saturation should be welcomed at the proper time. An analysis using steel and cement – two key materials for infrastructure – as indices predicts that in 2050 they will reach near saturation worldwide.

Saturation of man-made objects means the amount of materials that are newly required is equivalent to the amount of man-made objects that are discarded. If waste was recycled and used for new products, then the need for extracting natural resources would disappear. That is to say, at the very least, there would be no anxiety at least about metallic resources drying up, and they perhaps would no longer even be necessary in the first place. The form of society that we should have in the future is a completely circulating one that has minimized the volumes of both waste and resource mining.

1.2.4 Tripling Energy Efficiency

Setting the target value for energy efficiency starts having a theoretical value for total energy consumption. The difference between this theoretical value and actual energy consumption is the maximum possible value of energy-saving. Analyses of just how close we can approach this maximum value through technology is tantamount to a projection just how far we can push ahead with energy-saving.

For example, the theoretical value of energy consumption for transportation is zero. An object with a weight of zero requires no energy to travel. Accordingly, reducing the weight of automobiles and trains will be effective toward reducing power use, as well technologies that convert energy resources into an efficient force to power travel. I projected that by studying the possibilities of weight-reduction and drive technologies, it will be possible to cut energy consumption for transportation to one-quarter that of the benchmark year.

Heating demand ranks alongside automobiles as the largest source of energy consumption in the world. My thinking was that it will be possible reduce this to one-quarter by 2050 by improving the efficiency of a variety of factors in homes and offices by curtailing air-conditioning costs with improved heat insulation performance for buildings; reducing power consumption through the use of high-efficiency air conditioners, LED lighting, and the like; and improving efficiency on the power generation side of the equation.

Furthermore, efficiency increases will also continue in the areas of *monozukuri* (making things). Most metals oxidize in the air. Making metals from waste in a metallic state consumes markedly less energy than making it from natural resources in an oxidized state. In short, directing ourselves toward becoming a circulating society will also contribute to energy saving. Furthermore, the construction, home electronics, and other such industries could also cut their energy consumption in half. If such rates of reduction are averaged out to cover energy consumption for each of these categories, the amount that people consume as a whole could be kept down to less than one-third that of the benchmark year.

Thus, I came up with a projected amount based on a detailed study of each energy consumption item and a meticulous theoretical and technological analysis of those items that made the greatest contributions. I then took the resulting value, added the weights of the amounts of energy each consumed and averaged it to determine the appropriate amount a tripling of efficiency would be.

1.2.5 Doubling the Amount of Renewable Energy

The third point is to make it our objective that we double the amount of renewable energy. As of 1995, fossil resources accounted for an overwhelming amount of energy consumption at 80%. Renewable energy came next at 15%, comprising fire-wood at 10% and hydraulic power at 5%. Nuclear power also provided 5%. Accordingly, given that only 20% of the energy produced came from renewables that did not generate CO_2 including nuclear power, I set increasing that proportion to 40% through the use of renewable energy as the objective.

1.2.6 Increases in Both Comfort and Economic Performance

Each of the foregoing goals have been set rather high, but they certainly are not so reckless that they are unachievable. The discussions in the next chapter will show how progress has gradually been made toward achieving them in the two decades since 1995. Moreover, as we draw closer to a form of society we should have in the future, the quality of life will not decline. The amount of energy consumed can be kept down just by regularly replacing old air conditioners and refrigerators with new ones (Fig. 1.4). The convenience and ease of use of the new products should be better than before. Fuel consumption is not the only thing that improves with automobiles; their safety and comfort also improves. New enjoyments have arisen from choosing the car you want from among a variety of environmentally friendly options. Furthermore, the recycling rates for steel and concrete are rising in the industrial sector.

Are people creating inconveniences for themselves as the recycling rate rises, the efficiency of automobiles and home electronics improves, and the use of solar power spreads? Even if there are improvements to economic development, convenience, and a sense of safety, this will not be harmful. Furthermore, it will create new busi-

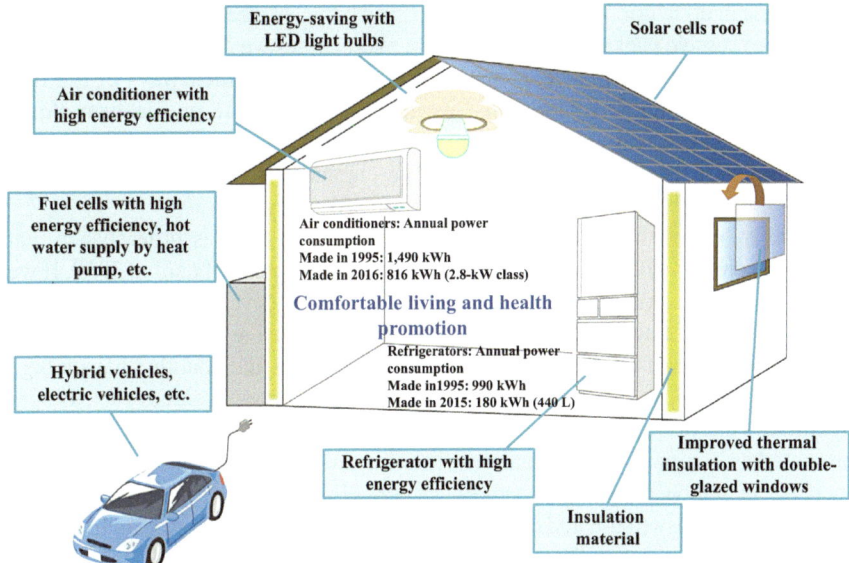

Fig. 1.4 Toward zero CO_2 emissions from houses by energy saving and creation. (Source: Created by the authors based on various materials)

ness opportunities such as the prospect of mega solar power. This is the true worth of Vision 2050.

A low-carbon society is one that keeps the use of the fossil resources that emit greenhouse gases under control. However, this does not mean the aim is to suppress human activities or economic development. A happy low-carbon society is not some pipe dream. Rather, it is a vision for feasible sustainable development.

1.2.7 Premises Consequent on Being a Realistic Vision

Vision 2050 is premised by growth in developing countries and standards of living being maintained in developed countries. If the people of developing countries have standards of living on par with those of developed countries in 2050, then the amount of energy consumed will swell to triple the level of the benchmark year. However, this does not mean that anyone has the right to say that for this reason developing countries want to maintain their statuses of living at current levels. The modernization of developing countries cannot be denied. Moreover, it is clear in the first place that if this happened, then we would be faced with a world of terrorism, coups d'état, and gloom.

Furthermore, while it is crucial that we rethink our lifestyles when it comes to reducing energy use, we must avoid being excessive in our expectations. There are things we can do in our daily activities like using bicycles when possible or tamping down on excessive packaging or not making photocopies we do not really need, but getting large groups of people to make extreme changes in their ways of life in a short period of time is difficult. The world does not operate on idealism alone. Vision 2050 also keeps reassessments of our lifestyles in mind, but it is limited in its expectations of any quantitative effects. The emphasis is on finding avenues for technologically improving energy consumption efficiency.

Another one of the premises of Vision 2050 is that there is little chance that renewable energy will have replaced fossil resources across the board in 2050. If we set aside hydroelectric power and the use of firewood in developing countries, then the percentage of total energy that renewables contributed as of 1999 stood at less than 1%. In contrast to fossil resources such as gasoline and coal that have a high energy density and can be used at any time, renewable energy sources such as sunlight and wind power have a sparse energy density and the volume of energy they produce has extreme time fluctuations. It is extremely difficult to come up with a scenario in which people will rapidly switch across the board from sources of energy that are easy to use and convenient to renewables that require planning when it comes to making use of them.

These are the premises that support Vision 2050. I undertook countless studies using the most up-to-date and detailed data I could acquire at the time I drafted it. Close to two decades have passed since it was formulated, but I still believe that everything in it to be current, including its premises.

The difference between developed and developing countries is likely to be a crucial factor that should be ascertained in the future. The amount of energy consumed in developed countries will definitely decrease. Their populations have already peaked out and are similarly shifting to a declining pattern. Japan's population peaked at 128,060,000 people in 2008. According a National Institute of Population and Social Security Research estimate, it might reach a maximum of 129,200,000 in 2050, or fall below 100,000,000, if the population decline accelerates. These population projections are based on suppositions about birthrates and average longevity, but assuming there are no major changes in conditions by 2100, it will drop to somewhere between 38 million and 65 million people.

When the size of the population falls, the total volume of energy consumed likewise falls. Furthermore, given that automobiles, houses, and household appliances will be even more energy efficient, the amount of energy they consume will drop even if we are still using them in the same ways. The amount of energy consumed in Japan rose continuously through 2004, but since then, it has been falling. Current levels of energy consumption are practically the same as those of the late 1990s when I worked out Vision 2050.

The amount of energy consumed drops as the population decreases and energy efficiency improves. Similar trends can be seen not only in Japan but also to varying degrees in all developed countries. It seems likely that developing countries will also navigate the same sorts of changes as development proceeds, their populations will reach their upper limits, and they will reach the saturation point with man-made objects. Vision 2050 was formulated as a realistic and rational vision statement.

Japan stands in the lead of the changes headed toward Vision 2050. This is because it is striving to move forward on becoming a happy low-carbon society in light of its population aging quickly, its birthrate being low, its population projected to shrink, and the country being faced with an enormous budget deficit. In short, in order to maintain affluence and the global environment, Japan is moving forward on the development of rational recycling systems, making energy use more efficient, and expanding the use of renewables as it strives to adopt new technologies and systems and improve productivity to make up for the decrease in population size. To achieve this, it is becoming increasingly necessary to have venues where hard-working and superior personnel can regularly acquire new knowledge, learn new ways of thinking, and develop new skills. People who have improved their abilities to cope with rapid changes can be made to skillfully recycle a variety of elements in the world around them. Doing so will enable a shift toward becoming a happy low-carbon society. It is incumbent upon us to design such a complex social system that includes all these elements, and then have the courage to put it into practice.

Note 1: In the present work, I use the expression "xxx tons of fossil resources." The "xxx" figure presented is a carbon conversion that takes into account the proportion of coal to petroleum to natural gas at the time. To express this in terms of CO_2 volume, multiply the number of tons of fossil resources by 3.1 to get the amount of CO_2. Please note that the proportions of coal, petroleum, and natural gas vary depending on the country and time period under discussion, so some deviations may arise.

Note 2: "Natural" energy (*shizen enerugī*) used as a technology used for sustaining the earth and "renewable" (*saisei kanō*) energy as used in the present work are the same thing. Recently, it has become commonplace to speak of "renewable energy" in English and its equivalent in Japanese for this kind of energy, so I follow that usage in the present volume.

Chapter 2
Progress on Vision 2050 Since 1995

2.1 Saturation of Man-Made Objects and the Material-Circulating System

2.1.1 Saturation of Population

"Saturation" is a basic concept that expresses the conditions of the twenty-first century. The world's population will also eventually reach saturation. Certainly, the U.N. projects the global population to continue growing in the twenty-first century. However, the root cause of this is the growth in average life-span. Population size is determined by the number of childbirths and average life-span, and the number of births in the world, excluding Africa, is already declining.

Each woman needs to give birth to two children—strictly speaking, 2.07 children due to the fact that more males are born than females—in order to sustain the population of a country. Few developed countries exceed that replacement rate of two, and those that do, do it just barely. Fertility rates continue to decline annually even in such countries categorized as newly industrialized like Mexico, Brazil, Thailand, and Indonesia. As of 2013, the rates for each stood at 2.19, 1.80, 1.40, and 2.34, respectively. That is, for populations outside of Africa, when the average life-span reaches saturation, they start to shrink.

In nearly every part of the world, the number of children born falls as the economy develops. The reasons for this are said to include a fading of the motivation to have children as a source of labor, an inability to withstand the burdens of having multiple children such as the costs incurred in providing their education, and people being more knowledgeable about such things as contraception. Furthermore, it is also known for a fact that as education – particularly that of women – becomes more widely available, it causes the fertility rate to drop.

Accordingly, if economic growth remains steady in Africa, and if education levels in particular make rapid strides there, then the number of births will reach

© The Author(s) 2018
H. Komiyama, K. Yamada, *New Vision 2050*, Science for Sustainable Societies,
https://doi.org/10.1007/978-4-431-56623-6_2

saturation or start to decline. The twenty-first century will be special era in the sense that the human population will be reaching its peak.

Vision 2050 hypothesized that the world's population will stand at 9.3 billion people in that year, and at present there are no major changes to that.

2.1.2   Saturation of Man-Made Objects

Man-made objects are being supplied and accumulated constantly in human society. The results of that accumulation are the forms of the modern city, but that is not to say this accumulation can go on without limit. The saturation when it comes to automobiles, buildings, and such in most developed countries is already striking.

Figure 2.1 shows how many residences and households there are in Japan. The country is believed to have reached saturation in this area around 2015. Japan has 50 million households, and around 60 million places of residence. That figure has gone well beyond one residence per household, with the number of residence exceeding that of households by more than 10 million. There are instances of people also having vacation or other second homes, but in any case, there are said to be 8 million houses that are vacant. In provincial cities, depopulation and an increase in the numbers of vacant homes whose ownership is unclear are becoming social problems. The annual demand corresponding to these 50 million households and 60 million residences is from 1.0 to 1.2 million residences when dividing it by the

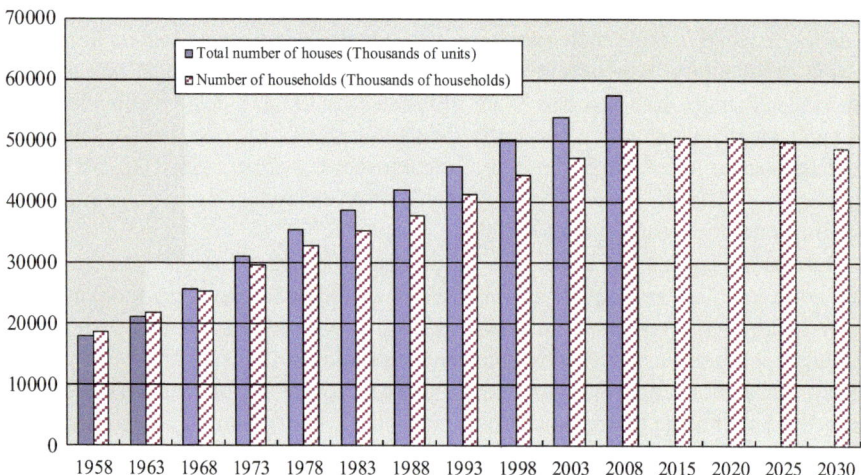

Fig. 2.1 Changes in the number of houses and the number of households in Japan. (Source: Created based on "2013 Housing and Land Survey" (Ministry of Internal Affairs and Communications) for 1958–2013, and "Household Projections for Japan (National Projections) (January 2013)" (National Institute of Population and Social Security Research) for 2013–2030)

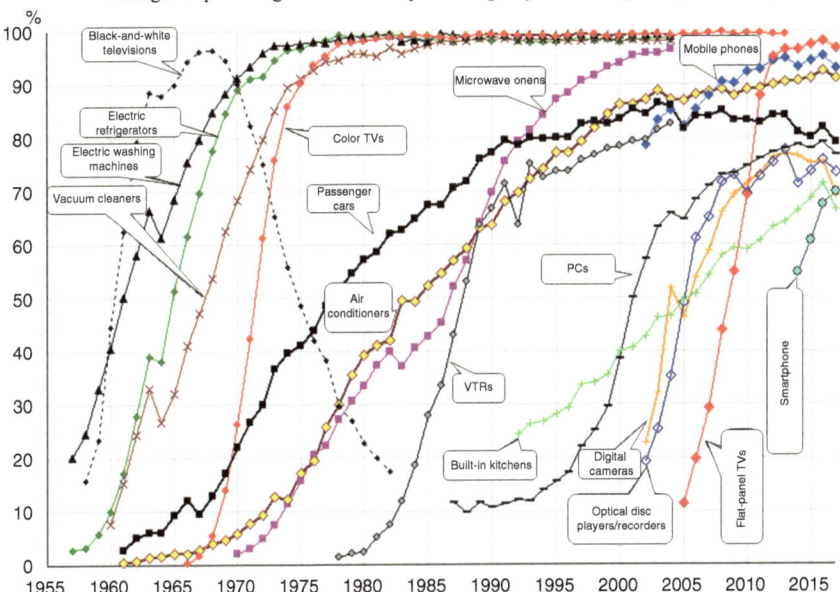

Changes in percentage of household possessing major durable goods (1957-2017)

Fig. 2.2 Percentage of households possessing major durable goods
(Data) "Consumer Confidence Survey," Cabinet office. (Source: Honkawa Data Tribune (http://www2.ttcn.ne.jp/honkawa/2280.html))
(Note) Applies to two-or-more-person households; applied until 1963 only to households in cities with a population of 50,000 or more; survey in 1957 conducted in September; surveys in 1958-1977 conducted in February; surveys in 1978 and beyond conducted in March; survey items changed from 2005; fall in many items in 2015 also due to effects of changes to the survey slip; digital cameras no longer include cell phones with built-in cameras; flat-panel TVs belong to the category of color TVs; optical disc players/recorders include those for DVDs and blue ray discs; from 2014, color TVs refer only to flat-panel TVs and no longer include to cathode-ray TVs)

expected 50-year service life of a home. A look at the number of new housing starts in recent years shows that it falls largely within this range.

Thinking about saturation of man-made objects in terms of what stage of economic development a developing country is at, we see it begins with infrastructure such as roads and railroads being laid down. As the country becomes a little more affluent, it extends to the high-tech products of daily life such as televisions and electronic goods. Finally, it reaches the expensive purchase that is the automobile. Japan has already reached saturation when it comes to most man-made objects (Fig. 2.2), and even China is steadily on track toward saturation (Fig. 2.3).

Figure 2.4 shows the number of automobiles sold per person in each country. For Japan, the number of cars owned is in the 60 million range, meaning there is basically one car for every two people. Similar conditions obtained in other developed countries like the U.S., UK, France, and Germany, with one car to two people representing a state of automobile saturation.

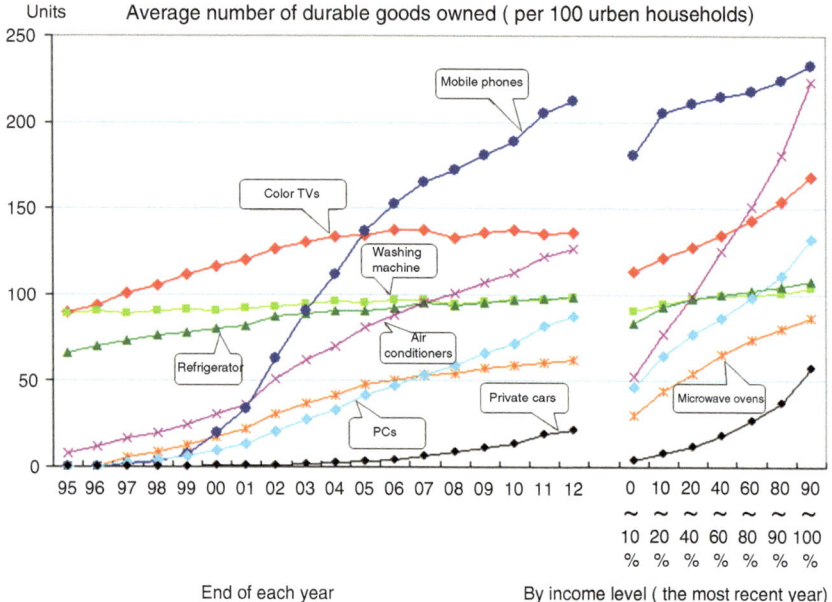

Fig. 2.3 Changes in percentage of households possessing major durable goods. (Source: "Honkawa Data Tribune" (http://www2.ttcn.ne.jp/honkawa/8200.html).

(Note) By sampling survey; farm households not included up to 2001, but all households became subject in 2002 and beyond; survey at the end of 2012

Source: China Statistical Yearbook

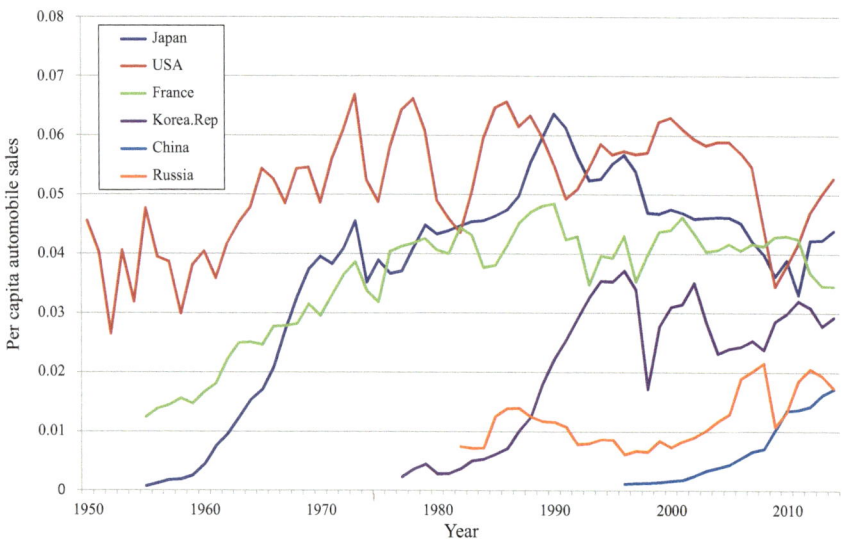

Fig. 2.4 Secular changes in per capita automobile sales. (Created by the Author. Source: Automobile sales according to Automotive Yearbook, population according to UNSD Demographic Statistics)

Even if it is at saturation, new cars can still be sold because there is demand for replacement or renewal. In the case of Japan, on average it takes about 12 years before an automobile is discarded, which means if we divided the figure of 60 million by 12 years we get 5 million vehicles disposed of annually. Since 1989, the number of new registrations has averaged from the 4 million to 5 million range. The number of new cars dovetails with the number that are discarded. Based on this, just how many cars can be sold per capita? There are 50 automobiles for every 100 persons. Fifty annually divided by 12 years is roughly four. So, in other words, the number of vehicles that are sold under the circumstances of automobile saturation is four for every 100 people, or in other words 0.04 vehicles per capita. A glance at Fig. 2.4 shows that developed countries are all converging on that figure. Furthermore, China is quickly closing in on developed country conditions. If China's development follows that of Japan and South Korea, then it will come close to saturation for automobiles in 5–10 years.

In short, saturation of man-made objects produces saturation in the number of units sold and ends up putting a halt to economic growth. This is what truly lies behind the low growth rates of developed countries. Accordingly, it would be best to think that no model for economic growth exists that extends from existing scenarios.

2.1.3 Saturated Demand for Substances and Materials: Cement

Being saturated with man-made objects also means being saturated with the resources used for man-made objects. Figure 2.5 shows the per capita production volume for cement at present around the world. Given that the exports and imports of cement are comparatively small, its production volume can approximate the amount brought into the country. Accordingly, the area below the line is the cement input volume for each country.

Cement is a good indicator for the state of a country's development and urbanization given that it used for roads, ports, dams, and similar projects. Every country eventually hits a peak in the increase of input volume needed to set up the large amounts of infrastructure en route to its period of so-called high-speed growth. In Japan's case, the volume rapidly increased starting in the 1960s to peak somewhere between 1980 and 2000, after which it has been in a period of decline.

The U.S. has been a driver for many years of contemporary civilization as represented by its car-oriented society. This is likely why it has grown at a sluggish pace through trial and error, rather than constructing paved roads and express highways all in one go. Its total sum input volume to date is approximately 17 tons per capita.

Meanwhile, since the 1990s, China has been moving forward with a mad burst of construction and has already reached the level of 22 tons per capita. In other words,

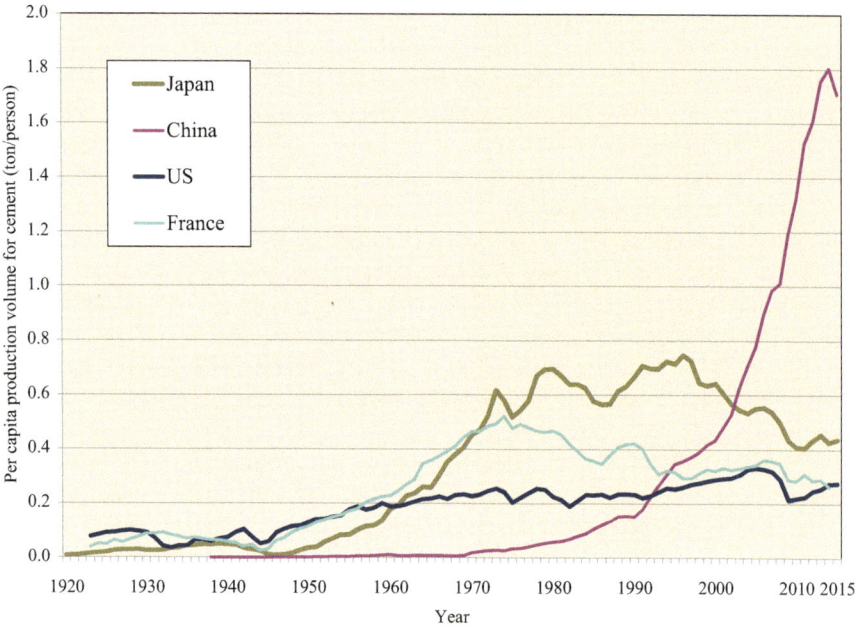

Fig. 2.5 Per capita cement production volume. (Created by the Author. Source: the global market. (Source: Yearbook, US Geological Survey), population (UN 2015 Revision of World Population Prospects))

with a population that is 4.5 times the size of the U.S., China has an infrastructure that is close to six times as big. Considering news reports that there are clusters of buildings that lack tenants, we cannot deny the possibility that it is already overabundant.

2.1.4   Saturated Demand for Substances and Materials: Iron

After cement, iron is the resource used in the greatest amounts. Its physical properties of strength and toughness can be controlled as desired by controlling composition and adding substances. Moreover, it is superb in that it can be recycled however many times one wants. Iron is already at saturation in Japan. The total input volume of iron associated with the input of new products is 30 million tons. The amount of iron including that from wasted man-made objects – in other words, scrap – is also 30 million tons. The fact that these two amounts are largely in equilibrium is as I discussed in the previous chapter. This corresponds also to almost completely saturated demand for man-made objects as shown in Fig. 2.2. Of course, new products such as smartphones will appear on the scene, but they are not so big that they will affect the balance in the total amount of iron. When it comes to saturated demand

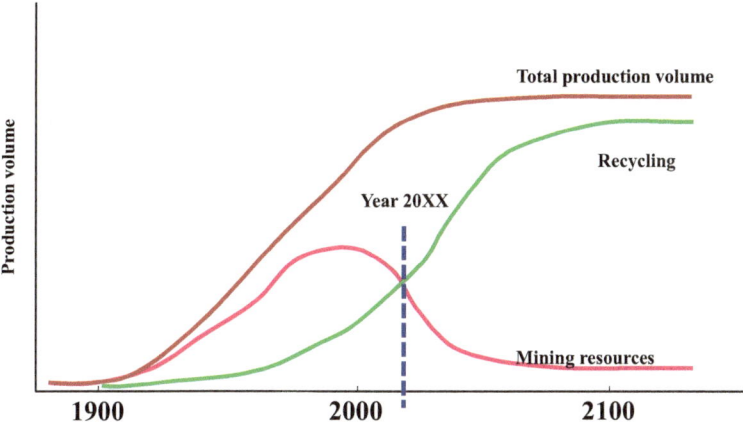

Fig. 2.6 Transition from mining resources for production raw materials to recycling

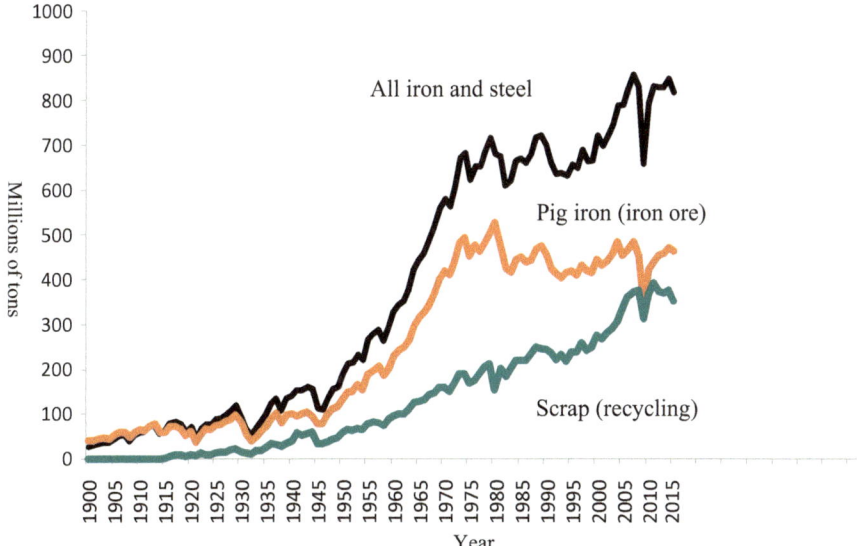

Fig. 2.7 Production volume of pig iron, scrap, and steel in the world excluding China. (Created by the Author. Source: U.S. Geological Survey Data Series, Steel Statistical Yearbook (World Steel Association))

for man-made objects that are large enough to affect the total amount, it is for objects at the final stage like automobiles and buildings. I have already spoken of automobiles and houses, but if you look at most of the cities in the world you can get a visceral sense that buildings are at saturation, too.

Figure 2.6 shows that production is shifting from resource extraction to waste recycling due to saturated demand for man-made objects as projected in Vision 2050, and gross production volume is reaching saturation. Figure 2.7 shows trends related to the corresponding amount of iron. Currently, China produces more than half the world's iron, and until recently most of the iron it produced was also consumed in China. Accordingly, the figure attempts to deduct China's contributions to the world total and look at global trends without China.

Comparing the two figures we see they are extremely similar. From this it can be seen that iron production has been gradually transitioning from the use of iron ore to the recycling of scrap. As of 2012, recycling already accounted for about half. This figure chiefly shows a state of saturation in developed countries. China is quickly reaching saturation, and eventually the whole world will reach that state and shift into an era of recycling.

Next, let us examine future trends with China included.

The cumulative amount in Japan is 1.4 billion tons. With a population of around 125 million people, that works out to 11 tons per person. That figure is the saturation value per person. If we think about its salience in terms of the population density of Japan and the density of man-made objects according to that, we can consider it to be the maximum saturation value per capita.

If we apply this 11 tons per person figure to China with its 1.3 billion population, we get a saturation value of 14 billion tons. The total amount of iron accumulated in China through 2012 is estimated to have been 8.9 billion tons. Annual demand is currently said to be 700 million tons, so the amount will reach 14 billion tons in 7–8 years. In proportion to population, this means it will reach saturation equal to Japan.

Next, let's look at the world as a whole. Currently, approximately 1.4 billion tons of iron are produced annually around the world. The portion of this that is iron produced from blast furnaces using iron ore amounts to 1 billion tons annually and represents the new cumulative amount globally. The remaining 400 million tons comprise recycled iron produced in electric furnaces using scrap.

The total amount of iron that humanity has produced and accumulated up to now is estimated to be about 30 billion tons. It exists in various places around the world as man-made objects. That constitutes an urban mine. Furthermore, if we assume that the amount of iron will continue to accumulate at 1 billion tons annually as it does today resulting in an accumulation of 34 billion tons by 2050, then the total size of urban mines will reach 64 billion tons. The saturated cumulative amount for a global population of 9.3 billion people at 11 tons per person is approximately 100 billion tons. Given that we will be at 64% of saturation volume in 2050, this is comparable to where Japan was around 1990. If the conditions that existed in Japan 25 years ago are produced around the world, then it is quite likely that demand for it will be at saturation.

Moreover, if we assume the life-span of man-made objects that use iron to be 30 years, then the cumulative amount of scrap produced in 1 year would be 1/30th the total or about 2.1 billion tons annually. Currently, scrap is being generated in amounts that far exceed the world's production volume. The amount of energy consumed in the future will consequently be more on the shift from scrap being markedly reduced to being thrown out than it will on the use of iron ore. That is to say, it may be assumed that a sense of saturation with iron will have permeated the world in 2050.

Under these conditions, if a material-circulating system for making all ironware from scrap is established, then humanity will be able to announce that it has finally gotten away from extracting iron ore.

To transition from extracting natural resources to making use of urban mines – this would be an effective step not only for iron but for all inanimate materials including cement, ceramics, aluminum, and precious metals.

This argument also holds largely true for biological materials like lumber and paper, but the fact that they are grown after they are harvested and after they have been recycled for use however many times they finally are burned and tidily used as energy constitutes a proper cycle. Plastics are also the same, but eventually by 2050, significant amounts of plastics will be made from biological resources such as wood.

Thus, while there are differences in the particulars of recycling methods based on the materials, the goal that humanity should be aiming for is a complete material-circulating system.

2.1.5 Hope for a Circulating Society

While people to date had seen saturation of man-made objects as a troubling issue from an economic perspective, it is a hope for humanity from the perspective of sustainability of resources and energy. First, amid all the concerns about the resource depletion, those about the limits for inanimate resources would go away. Those that remain would be over biological and energy resources, and as we will discuss later, these, too, would be fine rovided we proceed in accordance with Vision 2050 and the Platinum Society proposal.

Thus, when it comes to saturation of man- objects, we are moving forward as I projected two decades ago in *Chikyū jizoku no gijutsu*. In short, at present saturation is roughly being reached in developed countries. Because of he breakneck economic growth that China – a developing country at that time – has managed, it is closing in on saturation and may have even overshot it. Globally, 2050 will be the turning point on saturation of man-made objects, and it is conceivable that by that point, we should have put into place the technology, systems, and economy of a circulating society.

2.2 Energy Saving and Renewable Energy

2.2.1 Further Development Achieved in Energy Saving

Global energy consumption has increased by about 50% since 1995. However, Fig. 2.8 shows that energy consumption relative to GDP has declined 30%. Energy consumption relative to GDP is largely dependent on changes in industrial structure and technological improvements. Specifically, when manufacturing industries

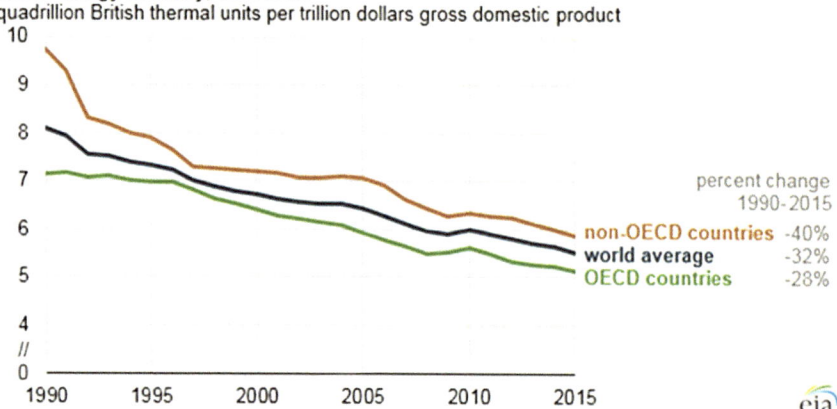

Fig. 2.8 Global energy-saving index. (Source: The U.S. Energy Information Administration (EIA))

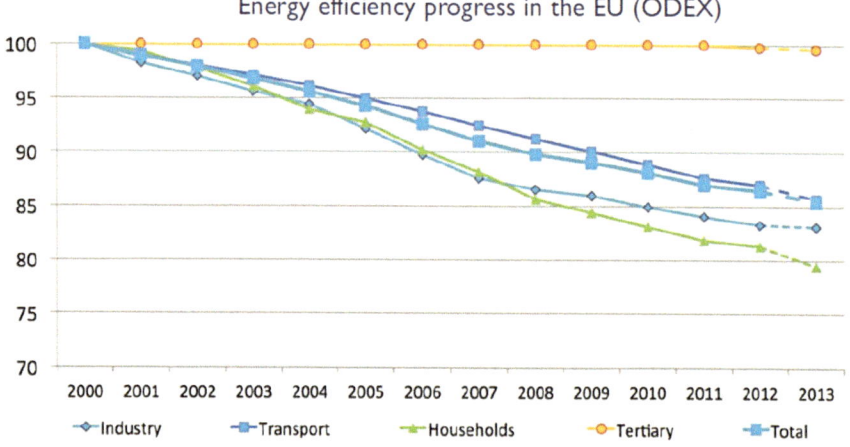

Fig. 2.9 Industry and energy consumption. (Source: ODYSSE)

shrink and service industries expand, energy consumption relative to GDP falls off and then goes down further due to improvements in energy technology. The energy-saving effects shown in Fig. 2.8 include both of these factors. In fact, an analysis of the period between 1995 and 2015 shows that the GDP increased by 78%, which outpaced the 50% rise in energy consumption.

Energy technology has also clearly improved. As can be seen in Fig. 2.9, from 2000 to 2012 energy consumption by energy-intensive industries dropped by as much as 15%.

Looking at automobiles, which account for 21% of energy consumption world-wide, we see that remarkable improvements have been made in the energy efficiency of leading-edge environmentally friendly vehicles. The Japanese automobile companies are global leaders in this area. Toyota, for example, has gone all out in its introduction of hybrid vehicles. Fig. 2.10 shows that energy consumption by new cars sold in the global marketplace also improved 20% from 2000 to 2012.

Based on the foregoing, it would be safe to say that we are on target for tripling energy efficiency by 2050 compared to 1990.

2.2.2 *Putting Renewable Energy at the Core of Energy Investments*

Figure 2.11 presents trends of electricity generation by source since 1995. Meanwhile, Table 2.1 shows the total energy supply and the share of each fuel from 1995 (when Vision 2050 was written) through 2015. Fossil fuels including oil, coal and natural gas have always accounted for the biggest share of overall energy production, while the share that they accounted for in electricity specifically was nearly constant at 65%, with composition for the remaining 35% exhibiting great changes. First, the absolute amount of hydropower increased by 1.6 times. Renewables other than hydropower – which at 1.4% was essentially close to zero – grew considerably, with wind power and solar reaching 6.4%. This growth has been based on market mechanisms, and renewable energy has already achieved price competitiveness. Since reaching its peak amount of electricity generated in 2006, nuclear power has declined slightly.

In terms of energy as a whole, fossil fuels account for 76.5%. The total amount of renewable energy including wind, hydroelectric, solar, geothermal, and biomass power grew by 1.6 times, and the share that renewables account for in the total energy supply has reached 18.8%. Meanwhile, nuclear power fell from 6.3% in 1995 to 4.8% in 2015.

As a result, when we add renewable energy to nuclear power, we see that the share of non-fossil fuel energy accounts for 23.6%.

The reason why renewable energy is being adopted is because costs have fallen. This is a result of a positive spiral of improvements in technology and an expansion of the market size, followed by further improvements in technology. As Figs. 2.12 and 2.13 shows, the costs of wind and solar power fell dramatically from the 1980s to the present, to 1/20th and 1/200th of their earlier levels, respectively. Even looking at just the past two decades since 1995, their costs have dropped to 1/4th and 1/20th of their earlier levels, respectively. Reflecting this state of affairs, renewable energy has been put at the core of energy investments.

We have long argued that such a fall in costs would occur. However, many people in Japan including energy specialists did not believe this would be the case. As a result, Japan is one step behind compared to the pace at which renewables are being adopted around the world. Japan lags far behind when it comes to wind power. The country was a pioneer globally with solar power in launching a national project

Comparison of fuel consumption by passenger cars
(new car sales)

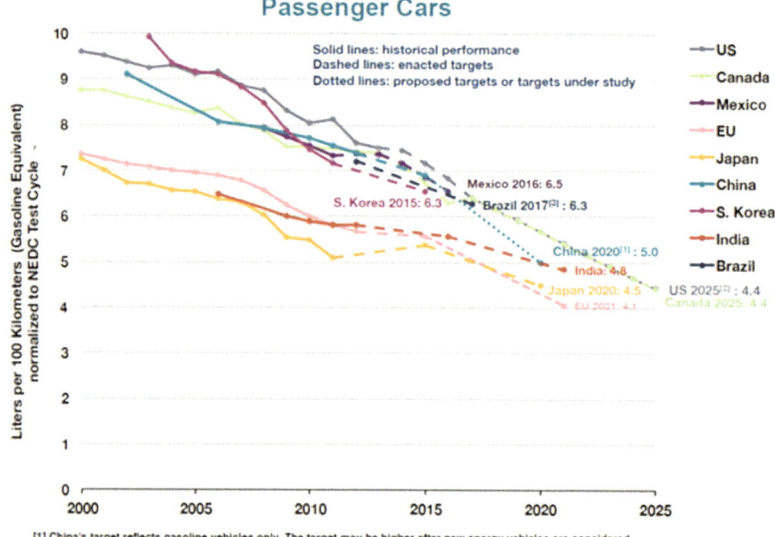

Source: ICCT(2014) Global Comparison of Passenger Car and Light-commercial Vehicle Fuel
Economy/GHG Emissions Standards, The International Council on Clean Transportation

Comparison of passenger car traveling fuel consumption

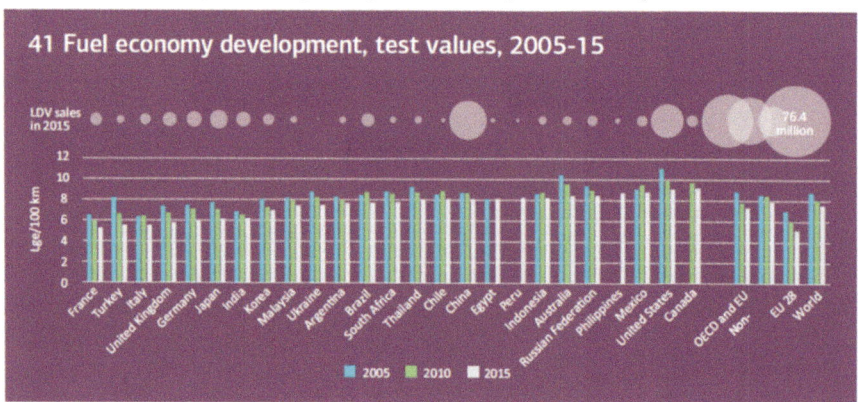

Fig. 2.10 Energy consumption of new cars sold in the global market. (Source: ICCT (2014)
Global Comparison of Passenger Car and Light-commercial Vehicle Fuel Economy/GHG
Emissions Standards, The International Council on Clean Transportation (above), Tracking Clean
Energy Progress 2017, IEA (below))

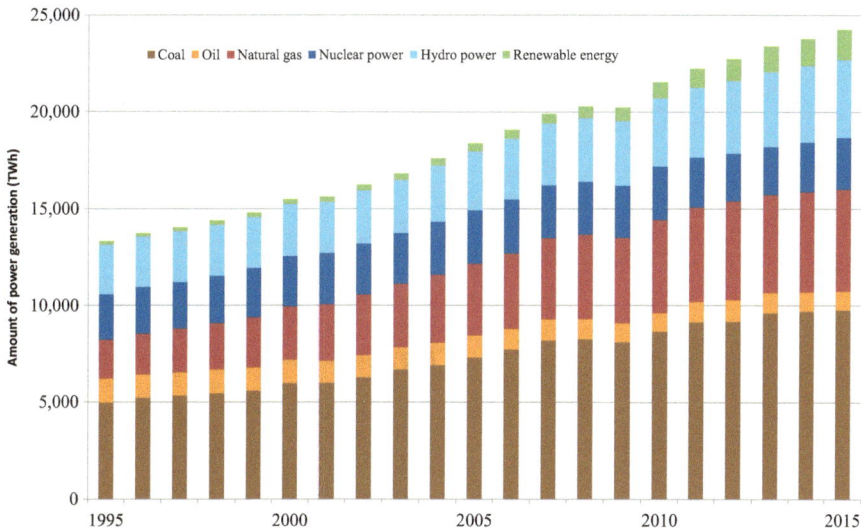

Fig. 2.11 Changes in power supply by resource in 1995 and beyond. (Source: IEA Energy Balances)

Table 2.1 Changes in total global primary energy supply

	Unit	Coal	Oil	Natural gas	Nuclear power	Hydropower	Firewood	New renewable energies	Total
1995	%	22.8	34.9	18.7	6.3	6.9	9.6	0.8	100
	Mtoe	2205	3372	1807	608	663	931	74	9660
2000	%	22.2	34.7	19.6	6.4	6.7	9.3	1.1	100
	Mtoe	2340	3660	2067	675	703	978	117	10,540
2005	%	24.5	33.3	19.5	6.0	6.5	8.1	2.1	100
	Mtoe	2947	4007	2352	721	786	978	249	12,041
2010	%	26.0	30.7	20.3	5.3	6.8	7.3	3.6	100
	Mtoe	3502	4131	2736	718	920	978	483	13,469
2015	%	26.8	29.4	20.3	4.7	7.2	6.7	4.9	100
	Mtoe	3918	4290	2970	689	1048	978	723	14,615

Note: New renewable energies refer to wind, solar, geothermal, etc.
Source: Created based on IEA Energy Balances; data for 2015 created based on World Energy Outlook 2016

called the Sunshine Project in 1974, and its technology clearly led the world. It fell behind in getting it out into the public at large, but it has adopted a feed-in tariff as it makes up for falling behind.

There has been no growth with nuclear power – a source of energy that. Like renewables. Does not emit CO_2. The amount of electricity generated at present is no more than 15% higher than the figure for 1995. In *Chikyū jizoku no gijutsu*, I wrote that owing to worries over safety, nuclear power would be an energy in a transition

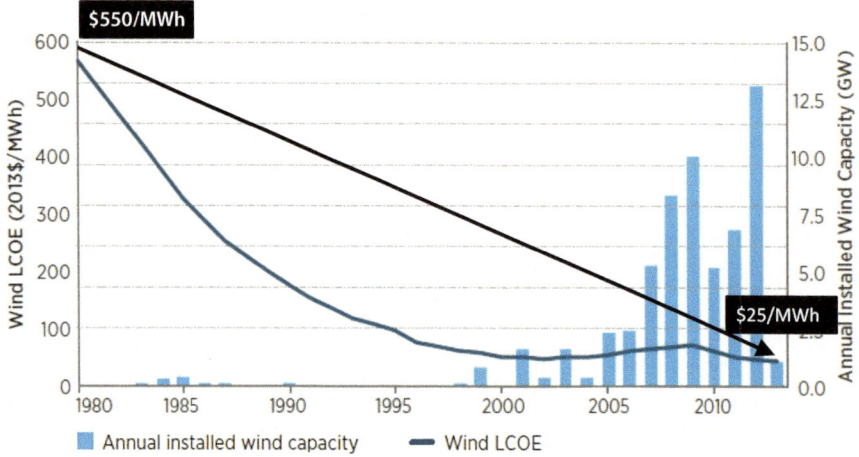

Fig. 2.12 The graph of "Wind Vision: A New Era for Wind Power in the United States, U.S. Department of Energy Wind Energy Technologies Office" has been processed. (Note: In the *Wind Vision*, good to excellent sites' are those with average wind speeds of 7.5 meters per second (m/s) or higher at hub height. LCOE estimates the PTC)

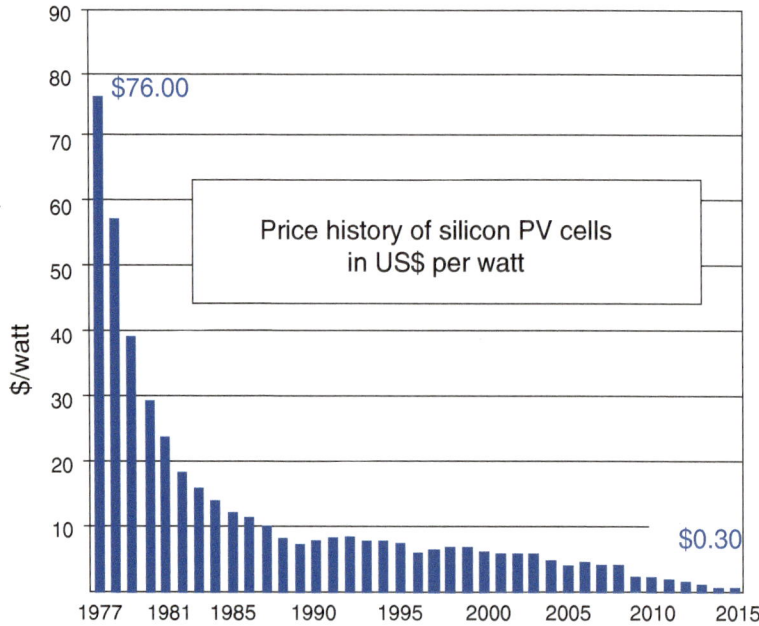

Fig. 2.13 Fall in solar power generation costs. (Source: Bloomberg New Energy Finance & pv.energytrend.com)

period from the mid-twentieth century to the mid-twenty-first century and account for around 5% of total energy. In fact, the reason for nuclear power's lack of growth is that anxieties over safety caused costs to rise and so it has no economic advantage. With the cost of renewable energy dropping dramatically, it is difficult to paint a scenario in which nuclear power will recover its cost competitiveness. It can be said that we were correct (in Vision 2050) to see nuclear power as an energy source going through a transitional period.

However, the challenge for Japan in debates over energy is the nuclear power problem. This is because it has reactors into which considerable capital investments were made in the past, and there are many people involved with them directly and indirectly. In other words, it boils down to the facts that exist now. The energy with which the sun bathes the Earth is more than 10,000 times as much as what human beings need right now. Renewable energy technology is what makes it easy and inexpensive to use that energy. Given that this technology is already economically competitive, it would not be an exaggeration to say that humanity has already settled on renewable energy for the future. Japan should quickly free itself from the yoke of nuclear power and push forward vigorously with energy efficiency and renewable.

By way of conclusion, the speed with which wind and solar power have made inroads exceeds even my projections at the time I drafted Vision 2050. Another miscalculation I made, in a good sense, was about the growth of hydropower generation. I thought that we might have already reached a limit on the construction of big dams, but in fact there was still room for development, to the extent that it might nearly double.

The significance of the doubling of renewable energy mentioned in Vision 2050 is that it called for raising the 20% share they had accounted for – 10% from old-fashioned biomass like firewood and manure; 5% from renewable energies such as hydro, wind, and solar power; and 5% from nuclear power – to 40%. That objective has not been achieved, but renewables alone now account for more than 12% where they had once been at 5%. In 2050, we will be in an era in which renewables account for more than half of the world's energy.

2.3 Vision 2050 as a Happy Vision

2.3.1 The Industrial Structure of Japan as a "Leading Country in Resolving Societal Problems" and Energy

Let us next discuss the major role that Japan has played globally in making progress on Vision 2050. Figure 2.14 shows the data on per capita GDP and energy consumption from 1965 to 2015 with the data for all three categories in 1973 designated as 1.

GDP and energy consumption both increased in unison until 1973. Those were the years of high-speed economic growth, with industry (especially the heavy and

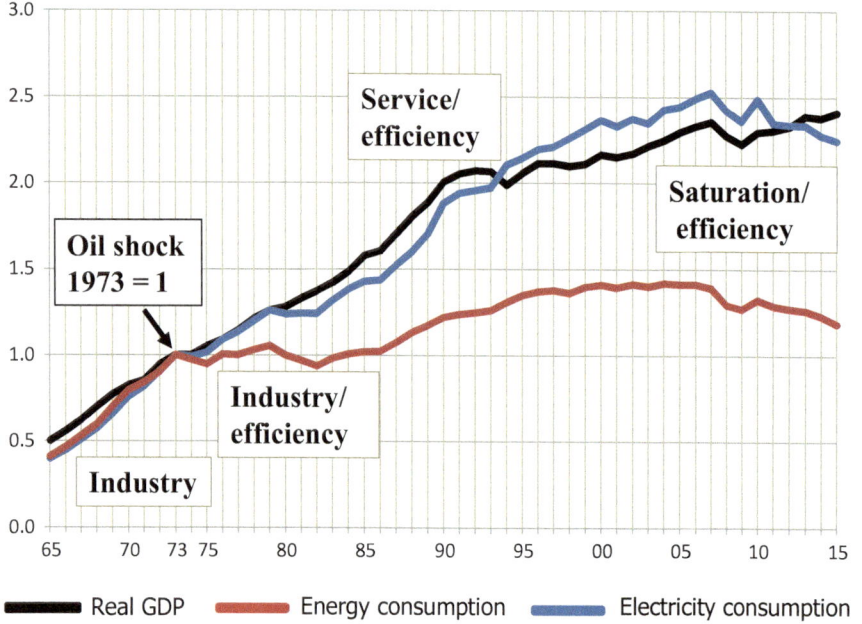

Fig. 2.14 Changes in GDP, energy consumption and electricity consumption. (Source: Created by the authors based on various materials. Real GDP ("National Accounts," Cabinet Office), electricity demand and final energy consumption ("Comprehensive Energy Statistics," Agency for Natural Resources and Energy) indexed by designating FY1973 as 1)

chemical industries) at the core. The first energy crisis occurred in 1973. The cost of petroleum rose sharply and all at once to 10–20 times its previous level. That dealt a direct blow to a global economy that was dependent on inexpensive petroleum, and delivered a heavy blow to Japanese industries in particular. However, Japan's core industries of the day – steel, chemicals, ceramics, and paper pulp – pushed forward vigorously on reducing power use and succeeded at converting a crisis into an opportunity. In other words, it succeeded in creating one of the best industries for "monozukuri" (making things) in the world, while simultaneously achieving economic growth without increasing energy consumption.

Incidentally, in 1973 the GDP stood at about ¥200 trillion and grew to ¥330 trillion by 1985, but energy consumption remained completely the same.

Subsequently, the core of economic growth switched from secondary to tertiary industries. Tertiary industries use about one-third the energy that secondary industries do to achieve the same GDP. This is why even as the economy grew, the amount of energy consumed was less than it had been prior to 1973. This was about the situation that existed in 1995 at the time when I drafted Vision 2050.

Energy consumption reached its peak in the years from 2000 to 2005, and since then it at long last has been falling. The phenomenon of energy consumption finally falling due to saturation of man-made objects and energy saving as laid out in the scenario for Vision 2050 is actually underway in Japan. In recent years, the economy has been growing only slightly, but nevertheless continues to grow. Comparing 2003 with 2015, we see that the GDP grew at an annual rate of 0.65% while energy consumption declined at an annual rate of 1.6%.

Electricity has presented a picture that is slightly different from that of energy as a whole. The work of service industries takes place mainly in offices rather than factories, and electricity is what it uses. The share of energy that electric power accounts for has been increasing. Currently, electricity accounts for 43% of energy consumption. As for demand, while the floor area of offices has continued to grow, it is gradually reaching a peak. Furthermore, energy saving at office buildings is making strides similar to those in the home. Under these conditions, while electricity use increased largely in unison with the GDP until around 2005, it peaked around 2006 and 2007 and then began to decline.

In Japan today, we are approaching saturated demand for those man-made objects that consume large amounts of energy like factories, automobiles, homes, and buildings, and when they are updated, we are improving their energy efficiency. As a result, Japan is entering an age in which energy consumption falls even as its economy grows.

Japan has been leading the world in translating into reality the task of how to address the issue of simultaneously achieving the economic growth required so that all the world's peoples can lead affluent lifestyles while dealing with the need to cut down on CO_2-emitting energy sources.

Thus, if we look comprehensively at the tracks that Japan has laid down to date, we see it has trod the path of a leading country in resolving societal problems aimed at in Vision 2050. However, it is at the mercy of the nuclear power problem, and its energy policy at present is such that it is difficult to see where it should put its focus. It should quickly free itself from the nuclear yoke, and continue to take the lead as it has already driven the world in the area of energy efficiency.

2.3.2   Certainly Japan Led the World

Figure 2.15 shows the current size of China's GDP and the amount of energy it consumes as well as future projects, as the IEA announced in 2016. Comparing this to Fig. 2.14, it is plain that China should adopt a course in the future that follows the course that Japan has already set down.

China's GDP and energy consumption have grown in the same way so far. This is the heavy and chemical industry sector-centered growth model that had prevailed in Japan until 1973. The GDP will grow by a factor of 3 times by 2040, but growth in energy consumption will stop at 30%. Japan has seen its GDP grow by 2.5 times from 1973 until now, but energy consumption grew only 24%. The fact is, China should do what Japan has been doing for the past 40 years, and likely will. However, this is a merely a matter of following the example of the past. Given that Japan is going to achieve further economic growth and reductions in energy consumption in

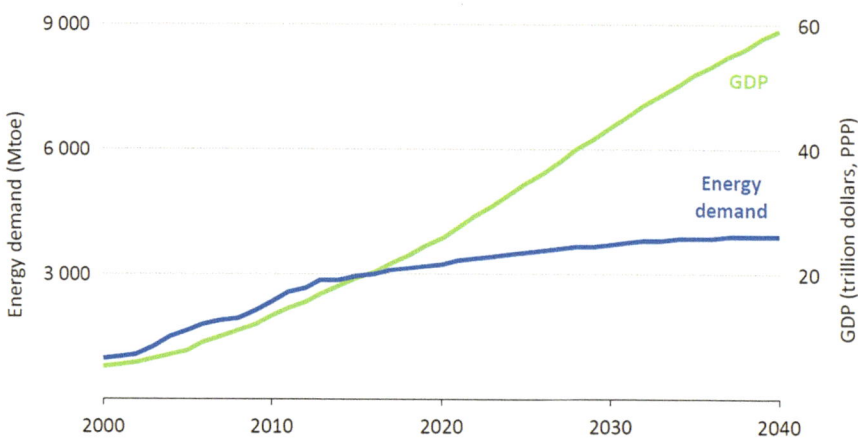

Fig. 2.15 China's GDP and energy consumption (according to projection by the IEA). (Source: World Energy Outlook 2015)

the future, China should parallel those moves and aim at keeping even bigger energy consumption under control.

Whatever the case, it is safe to say that Japan has presented the world with a superb model for managing both economic growth and energy saving.

2.3.3   The World Is Making Progress toward Achieving Vision 2050

Looking back at what has happened between 1995 and today, while it can be said that conditions are approaching those brought up in Vision 2050, there have been more than a few developments that I could not predict at the time of writing. For example, the U.S. – the world's largest consumer of energy – had been an energy importer at the time, but as a result of the shale gas revolution, it has become an energy exporter. Thanks to how easy it has become to make use of shale gas – which is present in abundant quantities – the price of gas has dropped in the U.S. Renewable energy has also been making inroads, while the strengthening of regulations on coal-fired power plants that should reduce dependence on coal continues. Many existing power plants have been shuttered or are planned to be closed, and nearly all plans for new ones have been cancelled. Of the approximately 500 utilities that had been sprinkled across the country, 180 have gone bankrupt.

As a result, the amount of CO_2 emitted in the U.S. is now declining. The total volume of emissions in 2015 had been reduced by 12% compared to 2005. The economy grew 15% during that period, and so like Japan, the U.S. managed to achieve both economic growth and CO_2 reduction. This situation could not have been predicted 20 years ago, but from the perspective of Vision 2050, it is a direction that should be largely welcomed.

Also, the speed of China's economic growth has surpassed expectations. Thinking about this by extension from the situation at the time 1995, the surmise was that China would not achieve saturation of man-made objects until further in the future than today. However, it has done so faster than expected, and what is more, if we think about things like the input volume of cement, it gives us the sense that it has even overshot that saturation. China has expanded beyond all measure when it comes to one shared yardstick. The situation perhaps is that the coastal and urban areas have gone beyond saturation of man-made objects, while it will still take a little more time for that to occur in the mountainous areas and provincial cities.

The concept of saturation of man-made objects was first put forth in Vision 2050. In recent years, observers mainly in Europe have begun to talk about the economy of the circulating society, using the term "circular economy." The circular economy argument calls for getting away from resource mining and greatly reducing energy consumption by combining recycling with saturation of man-made objects. This argument has not yet achieved the degree of logical consistency presented in Vision 2050. Furthermore, in 2000 the Japanese Government as a national policy adopted the three "R"s of "reduce, reuse, and recycle" through the Basic Law for Establishing a Recycling-based Society. Recycling comprises the re-use of something as a material and using combustion as an energy resource. The same value is assigned to material and thermal recycling. The three Rs in this sense comprise a fundamental way of thinking with which Japan should stand proud before the world.

Considered in this light, there is a strong sense of Japan's being a forerunner when it comes to the circulating economy. In particular, the concepts of the completely circulating society arising from saturation of man-made objects laid out in Vision 2050 and the energy-efficient society that comes with that are of a significance that even today should guide the world.

With regard to tripling energy efficiency compared to 1995, even today this is valid, including that "tripling" figure. That figure was a result of projecting just how far technology could close the gap between theory and reality in terms of energy consumption with respect to all of the bigger items in the amount of energy consumed and taking the weighted average of the projected figure. Various proposals had been offered before this such as Factor 4[1] and Factor 10[2], but these figures had thin theoretical foundations. A figure based on rational analysis would be quite reasonable. On the other hand, the objective of doubling the amount of renewable

---

[1] "Factor 4" is a concept first presented in 1992. It calls for using one-quarter of the resources and energy needed for products and services. This will make it possible to quadruple resource productivity (the amount of wealth and services that can be produced per unit of input volume of a resource). It aims to double wealth, while halving strains on the environment such as overuse of natural resources.

[2] "Factory 10" is a concept first presented in 1991. It argues that it is necessary to cut in half the amount of resources currently being used in the next 50 years in order to create a sustainable society. To accomplish this, it will be necessary for developed countries – which account for 20% of the world's population – to increase resource productivity (the amount of wealth and services that can be produced per unit of input volume of a resource) by a factor of 10.

energy requires adjustment. The objective should be to have it account for half of total energy by 2050.

The two most important issues that the Paris Agreement highlights are (1) energy-saving and (2) renewable energy. The part of Vision 2050 related to CO_2 is becoming something of which everyone in the world is aware. While there is no shortage of events that could not be predicted back in 1995, I believe it's safe to say that the three guidelines for action that provide its basic structure – "saturation of man-made objects and the circulating society," "energy efficiency," and "renewable energy" – in the not-so-distant future will start to receive recognition around the world.

Bibliography

Honkawa Data Tribune (2017) http://www2.ttcn.ne.jp/honkawa/2280.html. Accessed Jan 2018
Honkawa Data Tribune (2015) http://www2.ttcn.ne.jp/honkawa/8200.html. Accessed Jan 2018
ICCT (2014) Global comparison of passenger car and light-commercial vehicle fuel economy/ GHG emissions standards, p 8. https://www.theicct.org/sites/default/files/info-tools/ICCT_ PV_standard_Feb2014.pdf. Accessed Jan 2018
IEA (2017) Tracking clean energy progress 2017, p 103. https://www.iea.org/ publications/freepublications/publication/TrackingCleanEnergyProgress2017.pdf. Accessed Jan 2018
ODYSSEE-MURE (2016) Synthesis: energy efficiency trends and policies in the EU, p 11. http:// www.isi.fraunhofer.de/isi-wAssets/docs/x/en/projects/synthesis-energy-efficiency-trends-poli- cies.pdf. Accessed Jan 2018
The U.S. Energy Information Administration (EIA) (2015) World energy outlook 2015 launch pre- sentation. https://www.iea.org/media/publications/weo/WEO2015Presentation.pdf. Accessed Jan 2018
The U.S. Energy Information Administration (EIA) (2016) https://www.eia.gov/ todayinenergy/detail.php?id=27032. Accessed Jan 2018
U.S. Department of Energy Wind Energy Technologies Office (2015) Wind vision: A new era for wind power in the United States, p 30

Chapter 3
Technology to Support Low-Carbon Society (Using Energy)

3.1 Direction of Improvement in Energy Efficiency

3.1.1 "Daily Living" and "Monozukuri"

In a low-carbon society, we aim to suppress the use of energy that leads to CO_2 emissions. It is not a story about having to endure anything. As shown in Fig. 6, Japan succeeded in decoupling twice in history. A low carbon society is a concept that can coexist with economic growth and affluent life styles.

CO_2 emissions can be roughly divided into "daily living" and "manufacturing." Everyday life consists of three elements, "Transportation," "Business," and "Home." Manufacturing is "industry" itself.

Of the various fields, since transportation can be left alone and low-carbonization will still progress because automobile manufacturers are enthusiastic about improving fuel economy and developing eco-cars. Toyota launched a vision to reduce CO_2 emissions of new cars sold in 2050 by 90%. In addition, as new forms of services such as car sharing have emerged, a lifestyle without cars that has especially spread among young people in the city will also be a driving force for low carbonization.

Trends are similar in business and household energy consumption, and household electrical appliances such as lighting, air conditioning, hot water supply, kitchen each account for one-third. Initiatives to reduce energy consumption include replacement with high-efficiency products and introducing double-glazing windows with high thermal insulation effect. Although such updating of facilities is smoothest at the time of rebuilding, the rebuilding of houses is once every 40–50 years, and office buildings are once every several decades. Therefore, in order to achieve the vision in 2050, it is necessary to rationally promote renovation of existing buildings. For that purpose, we should establish a system like "paying electric bills as is" described later. For users, the big initial investment is the biggest factor of their hesitation about renovation. If organizations and groups such as funds that promote low carbonization shoulder the initial investment and users can direct

© The Author(s) 2018
H. Komiyama, K. Yamada, *New Vision 2050*, Science for Sustainable Societies,
https://doi.org/10.1007/978-4-431-56623-6_3

the electricity bill they saved from equipment renewal toward repayment, it would become easier for them to undertake renovations.

Meanwhile, in the case of monozukuri, the majority of energy consumption goes toward manufacturing materials. The industry with the largest energy consumption is the steel industry, followed by such industries as the chemical and ceramic, stone, and clay (cement) industries. When thinking about low carbonization of the manufacturing sector, it must not be forgotten that man-made objects such as iron and cement will eventually become saturated.

All man-made objects, whether it be an automobile or a building, are continuously introduced into the city, accumulated, and eventually become saturated. From now on, in order to build a new building in Tokyo, we have to destroy the existing building. That is, existing man-made objects are always discarded when newly introducing man-made objects. Recycling this waste is more energy efficient than digging and processing new natural resources. In the sense that waste from the city becomes a resource of new man-made objects, it is called an urban mine. It is also a word that has drawn attention in connection with rare metals, etc. in mobile phones. Monozukuri depends on whether it is possible to build a highly efficient material circulation system premised on the saturation of man-made objects and utilization of urban mines.

In this way, energy conservation that reduces energy consumption in individual sectors and expanding the use of renewable energy are two pillars for realizing a low-carbon society.

The IEA predicts that renewable energy will rank at the top of power generation by power source along with coal in 2030 and further increase in the future, with 60% of the investment in power directed at renewable energy by 2040. Meanwhile, 60% of investment in power in 2015 has already been directed at renewable energy, and this ratio reached 70% in 2016. Taking this into consideration, it should be all right to think that the introduction of renewable energy will probably progress at a faster pace than predicted by the IEA.

In this chapter, from the above viewpoint, we will consider technologies that support low carbonization.

3.2 Low Carbon Technologies in the Transportation Sector

3.2.1 Shipment Does Not Consume Energy?

Gasoline and electricity are necessary to move a car. Even when you ride a bicycle, you need a reasonable leg strength. Therefore, you tend to believe that energy is always consumed when you move something, but that is not accurate.

Actually, the theoretical limit of horizontal transportation energy is zero.

Imagine speed skating. When a skater starts skating on a skating rink, he/she kicks his/her skates powerfully with his/her legs. In order to generate kinetic energy, "work" is necessary, and a lot of energy is used here. However, when the speed becomes higher than a certain level, energy is theoretically not needed. The skater who reaches the goal is smoothly circling the rink while not kicking with his/her foot.

The skater will eventually stop by being caught by his/her coach, or grabbing the wall by himself/herself. At this time, kinetic energy turns into heat and is released into the atmosphere. The energy lost is equal to the energy used in starting to skate. Therefore, if you can store energy generated at the time of stoppage and use it at the start, you can continue to exercise forever.

A regenerative brake mounted on a hybrid vehicle (HV), etc. is exactly the application of this principle. The energy released when decelerating by applying the brake is saved in the battery and is utilized when starting and accelerating.

In the case of vertical motion as well as horizontal transport, the theoretical limit is zero. Pull the wire of the elevator on a pulley and attach a weight of the same weight to the other side of the elevator. If a good bearing is attached to the pulley, and without friction, no energy is needed to move the elevator up and down.

That is, friction is the cause of the energy consumed in transportation. A skater can continue skating by inertia after reaching a certain speed because the friction is small. Likewise, a satellite continues to fly and the Earth keeps going around the Sun because the friction is zero in outer space. However, there is friction in the real world. The bicycle stops unless the rider keeps rowing, and the skater cannot slide forever.

How can we reduce friction losses? This is the key to energy efficiency in transportation.

3.2.2 Energy-Efficient Cars Appear One after Another

Let's think about the energy efficiency of automobiles powered by gasoline engines that account for the majority of cars in the domestic market now.

A gasoline engine car burns gasoline in a cylinder, imparts a force to the cylinder head, rotates the shaft with that force, adjusts the direction and speed with many gears, etc., rotates the wheel, and runs. The overall picture is that the chemical energy of gasoline is converted to the work of a cylinder head, and that work is used to transport the car.

The chemical energy of gasoline is converted into work and heat. Since the law of conservation of applies here, chemical energy is converted by 100% if heat and work are combined. In theory, all gasoline may be converted to work, but only about 35% becomes work, and the remaining 65% is wastefully consumed as heat. Energy is thrown away in heat in various places, such as heat radiation from the exhaust gas and the engine, friction between the tires and the ground, and friction inside the car such as gears and transmissions.

Especially, when starting and accelerating, it requires a large amount of work, so a lot of frictional heat is generated and a considerable amount is released into the atmosphere. After reaching a certain speed, no energy is theoretically necessary,

Fig. 3.1 Latest model of Toyota Prius

while running at a constant speed, but in fact energy is consumed due to friction between the tires and the ground, and so on. If traveling at high speed, air resistance also occurs. Depressing the brakes during deceleration or stopping is friction itself, and even when stopping at an intersection, the engine is moving, and energy is used here as well.

Therefore, improving the following five points is a guideline for energy conservation in automobiles.

1. The efficiency of conversion from chemical energy to work is not 100%.
2. Friction of gears, etc. accompanying the transmission of force from the engine to the tires.
3. Friction between the tires and the ground.
4. Friction between the car body and air.
5. Friction due to braking.

One specific approach to improving efficiency is hybridization.

At the end of the 1990s when I proposed Vision 2050, Toyota's hybrid car "Prius" was the topic of conversation (Fig. 3.1). The Prius debuted in 1997 when the Third Conference of the Parties to the United Nations Framework Convention on Climate Change (COP 3) was held in Kyoto. The fuel consumption of the first model indicated on the car catalogue was 28 km/l (10–15 mode). At that time, the car of the same car class had fuel efficiency of less than 20 km/l, so it became a hot topic, as its fuel efficiency was outstanding.

The HV is a system that uses energy efficiently. It operates with an electric motor in normal cruising from vehicle start up to the mid-speed range. During deceleration and braking, it converts vehicle braking energy into a source of electric power. The engine stops not only when stopping at an intersection, etc. but also while driving, when the engine does not have to operate. In other words, it solves the problem of (1) conversion from gasoline to work and (5) friction due to braking.

Generally, HV is said to be suitable for urban areas because there are many signals, cars have to stop and go frequently, and there are many cases of starting with regenerative energy and electric power without operating the engine. When there is little regeneration during a long-distance drive, it consumes fuel just like a normal gasoline car.

Also, as the plug-in hybrid car (PHV) has a larger battery than the HV, it can be recharged not only by regenerative energy but also from an external power source. Therefore, it can run for several tens of kilometers without gasoline like electric vehicles (EV).

3.2.3 Car Energy Efficiency Increases Eightfold

How far can a car increase energy efficiency? Until a while ago, there were two choices, gasoline or diesel, but now a variety of cars are running on the market. EVs without engine are thought to be only light cars or compact cars, but recently, sedan type EVs or sports type EVs have appeared.

Fuel cell vehicles (FCV) have also become available on the market (Figs. 3.2 and 3.3). A FCV carries a fuel cell (FC) stack instead of an engine. The mechanism of FC was discovered in the early nineteenth century, and its history is old. For application to automobiles, it was a challenge to install gaseous hydrogen at room temperature on cars, but a 70 megapascal high-pressure hydrogen tank using carbon fiber reinforced plastic was successfully commercialized. A FCV generates electricity through the reaction of hydrogen and oxygen in the air, and is propelled by that energy. CO_2 emissions during driving are zero as in the case of EVs, but a FCV has more power than an EV, better driving performance, and longer cruising range.

Figure 3.4 plots a commercial model with fuel consumption on the vertical axis and vehicle body weight on the horizontal axis. It is worth noting that in Japan, we use the mileage per 1 l of fuel to represent fuel economy, but in this figure, the

Fig. 3.2 Toyota FCV "Mirai"

Fig. 3.3 Honda FCV "Clarity Fuel Cell"

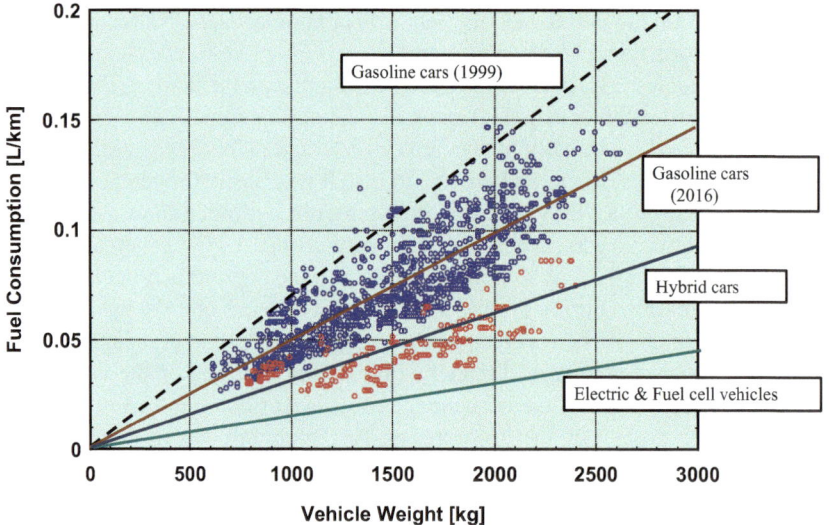

Fig. 3.4 Structuring of Knowledge: Energy efficiency of automobiles increases to $2 \times 4 = 8$ times. (Created by the Author. Source: carview!)

Western style of fuel consumption per kilometer is used. That is, since the vertical axis is the fuel consumption per unit distance, if the same weight is compared, the larger the value of the vertical axis, the worse the fuel consumption.

When other conditions are constant, fuel consumption and body weight are proportional. Therefore, the data of each model is plotted in general on a straight line, and the fuel consumption amount approaches zero as the weight of the vehicle body gets lighter. In theory, a car can move without consuming energy if the body weight is zero.

Indeed, energy-saving races are held all over the world. Drivers compete on how much they can reduce fuel consumption, or how far they can travel on 1 l of gasoline. A few years ago, Guinness record was set by a driver traveling more than 5000 km on 1 l of fuel. At that time, the weight of the car was only 25 kg. Even with the driver's weight of 45 kg combined, the load was 70 kg.

Since commercial vehicles must also provide safety and comfort, it will be difficult to reduce weight this far, but considering efficiency, it goes without saying that the lighter the car body, the better. Besides downsizing the car body, using light materials is a weight reduction method. In super cars and luxury cars, materials effective for weight reduction of the car body while maintaining a certain degree of robustness are being actively used, such as aluminum, carbon fiber reinforced plastic, and high-tensile steel, which is an alloy of iron.

Well, let's return to Fig. 3.4. Comparing the fuel consumption of all cars in 1999 with that of cars in 2016, the current consumption is 33% less for the same weight. This is an energy-saving effect due to progress in driving power technology over the past 17 years. As the weight of cars of the same size is becoming lighter through further weight reduction, it is considered that the fuel consumption of the same type of cars has been reduced by about 40%.

While the fuel consumption of HVs is less than half of that of 1999 gasoline engine cars, that of EVs and FCVs are even half of that of HVs. In other words, the energy efficiency of HVs is twice as high and that of EVs and FCVs is four times as high as that of gasoline engine vehicles.

In the future, with technological innovation, the car body weight can be made lighter. If we can reduce the weight of a car by half, energy efficiency will be doubled so we can make EVs and FCVs eight times more efficient.

In the discussion on energy, it is important as to what electricity and hydrogen are made from. Here, I want to note that electricity is calculated based on Japan's standard power supply configuration, and for hydrogen as well, it is calculated based on the value derived when water is subject to electrolysis under the standard power supply configuration.

3.2.4 A Rich Car Life with Diverse Eco Cars

Prius has carried out three full model changes so far, and each time, Toyota has developed a new hybrid system to optimize energy efficiency. Hybrid technology has been polished over the past two decades, and I think that citizens' understanding of HV has deepened.

Initially at the time of release, the Prius had a higher car body price compared with the same-class cars, and there were also opinions questioning, "Are you really getting it?" Recently, however, discussions on simple loss and gain hid behind the scenes. Certainly, the initial cost may be high, but if used appropriately, fuel consumption can be suppressed, and in the long run, it is economically advantageous. Less fuel consumption means less environmental impact. It can be said that there are more consumers who are making wise choices in terms of comprehensive satisfaction rather than because these cars are becoming cheaper.

Not only can EVs and FCVs that do not have an engine contribute to a low-carbon society but also they present no concerns about noise and exhaust gases that were associated with engine vehicles, so they are friendly to passengers and people

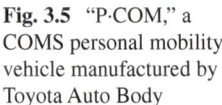
Fig. 3.5 "P·COM," a
COMS personal mobility
vehicle manufactured by
Toyota Auto Body

outside the car. In addition, they are compatible with new technologies such as a driving support system including automatic braking and automatic driving systems because they are electrical drive vehicles, and therefore, they can be driven safely and comfortably.

Since a variety of models are coming out, it is fun to be able to choose according to lifestyle. Some owners are saying, "An EV is sufficient because I mainly use if for shopping in the neighborhood." Still others are saying, "Sometimes I go on long drives, so a PHV is better."

In rural cities, cars are the main means of transportation. Because cars are indispensable even for commuting and a bit of shopping, the trend has been toward one person owning one car, but it is a waste that only one driver is on board a minivan or sedan. Recently, the development of one-seater EVs called city commuters or personal mobility has been proceeding (Fig. 3.5). Because they are compact and the specs are suppressed, the price is also cheaper compared to EVs in general. Smart ways of using these vehicles will become possible: small EVs for everyday short distance travel and PHVs for going out on a drive with family and friends.

Or, it may not be surprising to see people who possess ordinary cars and small EVs like those who have separate uses for cars and motorcycles. Even if you own more than one, you cannot move at the same time. As mentioned above, the energy efficiency of individual cars improves, so you can enjoy the pleasure of riding various cars without imposing environmental burden. With regard to long-distance-drive cars that are not used frequently, it is also a wise choice to use them for car sharing or car rental.

Sharing business has spread to various fields as a result of the progress of IT. Ideas such as sharing personal mobility in the area and lending it to tourists have also emerged, and from the viewpoint of regional revitalization, the evolution of cars is receiving a lot of attention.

In order to promote the low carbonization of the transportation sector as a whole, it is important to increase the energy efficiency of each car, and at the same time, to utilize it at the right place according to the characteristics of each car. The emergence of a variety of mobility has brought about the joy of making choices, and new business opportunities are being created. CO_2 emission reduction and affluence are concepts that can fully coexist.

3.2.5 Modal Shift in Movement

Private passenger cars account for 47.5% of CO_2 emissions in the transportation sector. This is followed by cargo vehicles and public transportation (buses, taxis, railways, ships, and airlines), which respectively account for 35.1% and 17.4% of the CO_2 emissions (Fig. 3.6). Therefore, in order to realize a low-carbon society, it is necessary to promote reduction of CO_2 emissions by cargo vehicles.

A cargo vehicle refers to a truck, and there are commercial cargo vehicles used by shipping companies and a private cargo vehicles owned by non-transport operators such as farmers or shops. When comparing the CO_2 emissions per ton-kilometer, which is the product of the transport weight (tons) multiplied by the transport

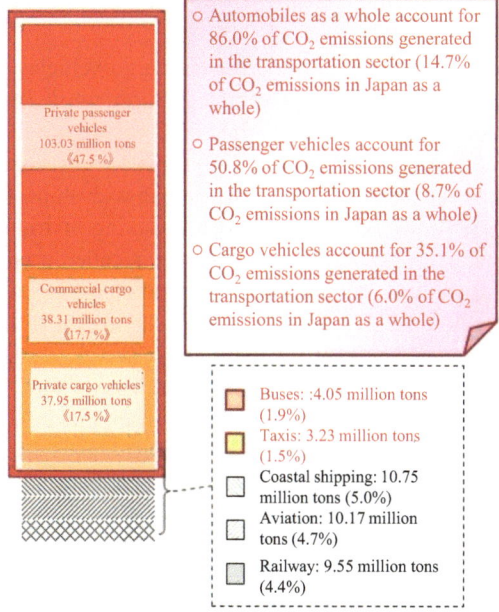

* Due to rounding off, there may be cases where the numerical values of the total do not match.
* Created by the MLIT Environmental Policy Division from "Japan's Greenhouse Gas Emissions Data (FY1990-FY2014) Final Figures" released by the Greenhouse Gas Inventory Office of Japan

Fig. 3.6 Breakdown of CO_2 emissions in the transportation sector. (Source: Ministry of Land, Infrastructure, Transport, and Tourism. http://www.mlit.go.jp/sogoseisaku/environment/sosei_environment_tk_000007.html)

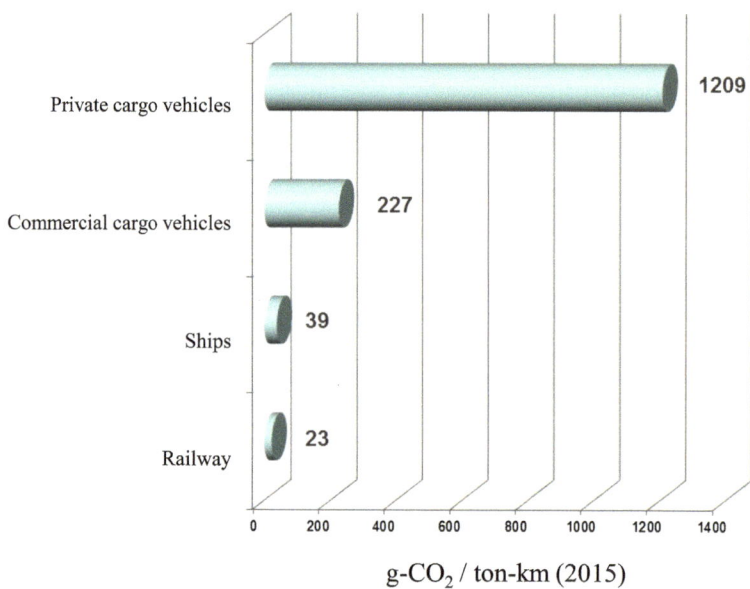

Fig. 3.7 Amount of CO_2 emissions per transport volume (cargo). (Source: Ministry of Land, Infrastructure, Transport, and Tourism)

distance (kilometers), as shown in Fig. 3.7, private cargo vehicles emit about six times more CO_2 than commercial cargo vehicles. Furthermore, the commercial cargo vehicles emit more than five times as much CO_2 as ships and eight times as much CO_2 as railways. Since the amount of CO_2 emissions is smaller with ships than with cargo vehicles, and smaller with railways than with ships, switching from truck transport to ships or railways can contribute to reducing environmental impact.

Changing the means of transportation in this way is called a modal shift. The Ministry of Land, Infrastructure, and Transport has promoted a modal shift for more than 10 years, but it has not progressed sufficiently. The reason is that truck transport is convenient. You can go anywhere at any time on a truck, and you can deliver in small quantities. On the other hand, ships and railways have low degree of freedom in operation routes and time schedules, and when picking up cargoes and delivering them to their final destinations, trucks are eventually used in combination. If the shift is not done on a reasonable scale, cost benefits are hard to come by.

However, the logistics industry is currently faced with the big problem of shortage of truck drivers. The aging of drivers is progressing, but there are no young volunteers because of the severe labor environment such as long working hours and low wages. Even if salaries and conditions are slightly improved, there are reportedly no responses to job advertisements. On the contrary, the number of logistics/ home delivery is increasing due to such developments as the rise of online shopping. If the load per driver increases, it is clear that the risk of delays in delivery and traffic accidents will increase, but in the long term there is no prospect of resolving the shortage of manpower. The logistics industry is now being forced to build a new business model.

Modal shift can contribute to solving such problems in the logistics industry. If you replace a truck with the railway, you will not have to secure a driver for that section. If companies dealing in small-lot delivery work together well, it is possible to secure the benefits of scale that can be felt using railways and ships such as increasing the loading efficiency of containers and securing cargo on the return trip.

A technology that has emerged that can compensate for shortage of human resources is automatic driving. Isuzu Motors and Hino Motors announced that they will jointly develop truck and bus automatic driving systems. They are aiming at convoy driving consisting of three or more trucks. During the experiment, a driver will ride on each of the vehicles, but finally, there will be a driver only on the leading vehicle, and the following vehicles are planned to be unmanned. The vehicles are not physically connected, but the idea is like that of a one-manned train. This makes it possible to increase transportation efficiency while reducing personnel expenses.

In addition, if complete automatic operation is realized in the future, the driver will be released from the obligation to monitor the forward direction, so it will be possible for the driver to engage in processing vouchers, sales activities, product planning, etc. inside the vehicle. Speaking of truck drivers, it is said to be a typical example of physical work, but it may be transformed into creative work as a result of progress in technology.

3.3 Low Carbon Technologies in the Home and Business Sectors

3.3.1 Promotion of Energy Saving Is Economically Advantageous

Where is energy consumed in the household sector? Figure 3.8 shows the percentage of energy consumption by usage in the household. The largest proportion is power, lighting and others, followed by hot water supply and heating. Power means energy consumption by electric appliances such as refrigerators. There is no big difference in this ratio even in the office, so with regard to low carbonization in the home and business sectors, the key is how much consumption can be reduced in these large energy consumption sectors.

For example, refrigerators and air conditioners will be replaced with the latest models with high efficiency, and lighting will be replaced with high efficiency LED bulbs. With regard to hot water supply, energy waste is reduced by simultaneously obtaining hot water and electricity through the use of home fuel cell energy farms and the heat pump electric water heater EcoCute. If heat insulation is installed firmly on the walls and floors and double glazing with excellent heat insulation performance is installed for the windows, the heating and cooling efficiency of houses and buildings will be improved. By doing all that we can do in this way and expecting future technological innovation, we can reduce the current home energy

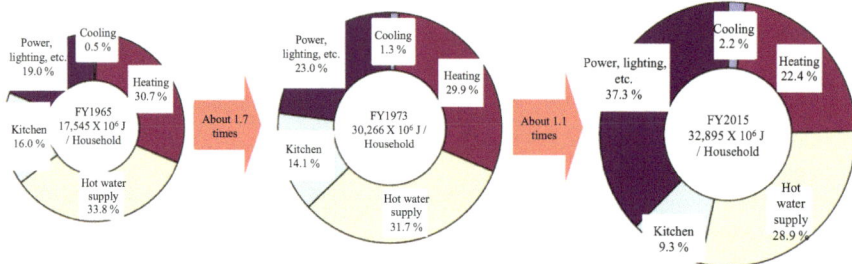

Fig. 3.8 Changes in unit energy consumption intensity per household and energy consumption by use. (Source: Energy White Paper 2017)

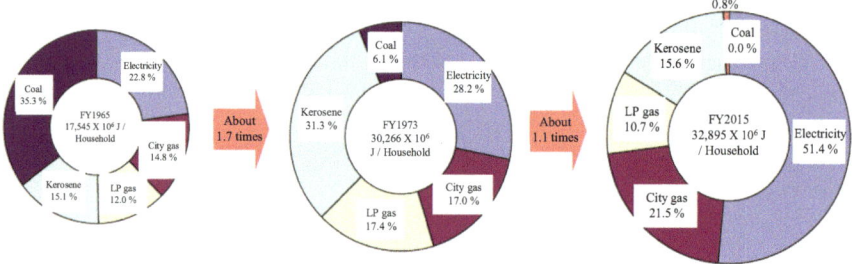

Fig. 3.9 Changes in consumption by energy sources in the household sector. (Source: Energy White Paper 2017)

consumption by a quarter in 2030. Then, if the remaining quarter of the energy can be covered by power generation by solar cells, etc., the energy brought in from the outside would be zero. If the amount of electricity generated exceeds consumption, it is also possible to sell energy.

If we list energy saving measures such as these, some people say, "Energy conservation cannot be done because it costs money." However, it does not mean that we have to do everything right now.

Even if you use electric appliances such as air conditioners and refrigerators quite normally, they will be replaced every 10 years. It is impossible that newly purchased products are less energy efficient than the models 10 years ago. Low carbonization is possible by simply replacing the older models.

Energy efficiency is also improved in recent models of Enefarm and EcoCute. Besides, as the prices have dropped considerably since their launch, it should be easy to recognize the cost benefits of introduction. Insulation materials and double-glazing windows are somewhat difficult to incorporate into existing houses, and the costs are relatively high. However, the energy saving effect of heat insulation is great. In case of new construction or large-scale renovation, installation should be considered positively (Fig. 3.9).

3.3.2 *Energy Conservation Will Be a Business Opportunity*

Investments in energy saving are recoverable. If energy consumption decreases, payment of electricity charges and gas fees will be reduced as a matter of course. If energy consumption is reduced to one-quarter, the amount of pay-per-use excluding basic charges will also be one-quarter. Even if the electricity fee rises and doubles in the future, it will cost only one-half of the current price. From a long-term perspective, it will be profitable to promote energy conservation.

This means that there is a business opportunity in promoting energy conservation.

The Center for Low Carbon Society Strategy (LCS) and The University of Tokyo have jointly proposed a mechanism of "paying electricity as is." This is a mechanism to ease the burden of households in introducing measures for reducing carbon emissions, aimed at making the energy saving/renewable energy of households significantly progress without relying on subsidies.

For example, when installing solar power generation systems and household fuel cells, financial institutions will finance initial costs and electricity fees saved by their introduction will be used for repayment. As a result, the initial investment borne by the household will be zero. As monthly payments will not increase unlike monthly installments, equipment for low carbonization can be introduced even in homes that lack sufficient funds. It is advantageous for society as a whole to be able to promote low carbonization measures without subsidizing it.

However, someone needs to take over the initial cost. In the UK, a non-profit enterprise was established and investments are made mainly by the UK Ministry of Energy and Climate Change. In Japan, cases of solar power generation promotion by a subsidiary of Osaka Gas are well known, but since credit screening takes time, it is not suitable for small scale projects. If households are being targeted, it is still appropriate to establish a fund. In addition, detailed adjustments are necessary to produce truly fruitful measures. The results of verification tests conducted at five places such as Nagaizumi-cho in Shizuoka Prefecture and Shimokawa-cho in Hokkaido from 2015 are awaited.

Meanwhile, the ESCO (Energy Service Company) project is already established as a business. The trustee of the ESCO project offers consultations on low carbon measures for office buildings, provides comprehensive services, and obtains compensation from the water utility cost that could be reduced. After the contract period has ended, the amount of reduction will be the profit that the consignor will be making.

When there are major events affecting the economy, such as the collapse of Lehman Brothers and the Great East Japan Earthquake, investment tends to be narrowed down, but investment in energy saving can certainly be recovered later. It is hoped that investors such as pension management organizations that ought to firmly recover their investments will become aware of this quickly.

3.3.3 Household Energy Consumption Is Consolidated into Electricity

Looking at energy consumption in the household sector by energy source, coal accounted for the largest proportion in 1965, followed by electricity, kerosene, and gas. In the 1970s, the proportion of coal used for heating, boiling, and cooking sharply decreased as lifestyles became modernized. In place of that, the ratio of kerosene increased significantly. The proportion of electricity and gas also increased.

In 2014, the proportion of kerosene decreased, and electricity came to account for the majority. The means to get warm has changed significantly from stoves that directly burn coal and oil to heating appliances that use electricity such as hot carpets and air conditioners. Even in the kitchen, there are more households that use IH cooking heaters, rather than gas.

When comparing heating efficiency from fossil resource consumption, air conditioners have the highest efficiency. Heat pump technology is used for air conditioners. The mechanism will be described later in detail, but the heat pump can draw heat from the outside up to six times the electricity consumed and supply it to the room. Since the conversion efficiency from oil to electricity is about 40%, it has a heating effect that is more than twice the heat from combustion of petroleum.

Electric heaters are heating devices that similarly use electricity, but they are less efficient than oil stoves. It is because electricity has already abandoned 60% of fossil resources in the power plant in the form of heat, and the remaining 40% will be directly turned into heat indoors. A heat pump also converts electricity to heat, but since it pumps up heat many times than that, it is highly efficient. Moreover, the air conditioner has a lower risk of fire than the oil stove, and the indoor air is not going to be polluted, so it is convenient for the consumer.

The energy used in the household sector will continue to increase in the future. Electricity can be made from petroleum or gas, or it can be made from renewable energy such as solar or wind power. Promotion of low carbonization in the household and business sectors shall be considered based on the premise that the energy to be used is consolidated into electricity.

3.3.4 Eco Houses Are Also Friendly to Their Occupants

Japan has created various energy-saving products and pursued energy efficiency, but energy saving measures have not sufficiently advanced for buildings. Even with existing buildings, the introduction of double glazing and heat insulation enhances thermal insulation property and air tightness, and the efficiency of energy use is dramatically improved. Zero-energy housing that is comfortable without introducing energy from the outside is one ideal form.

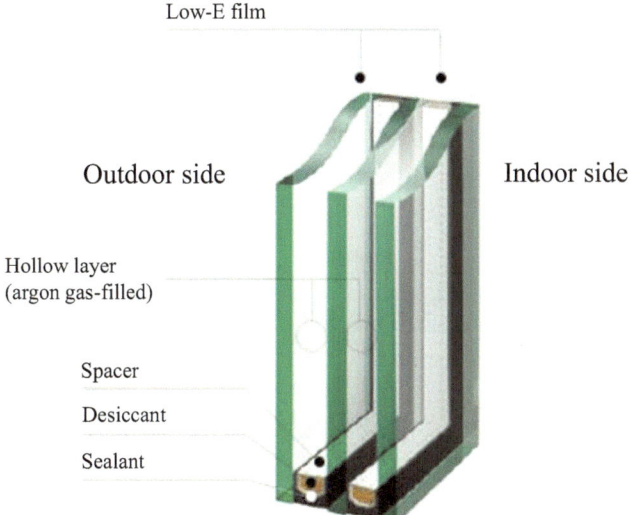

Low-E film

Outdoor side   Indoor side

Hollow layer
(argon gas-filled)

Spacer

Desiccant

Sealant

Fig. 3.10 Structure of triple glass. (Courtesy: AGC Asahi Glass)

Double glazing glass that is indispensable for the realization of the zero-energy housing is PairGlass, which consists of several sheets of glasses stacked together. Low-E double glazing glass with thin film coating of silver on the hollow layer side of the glass has been attracting attention for building use. AGC Asahi Glass provides multiple line-up of Low-E double glazing. In the area west of the Kanto region, solar radiation shielding type glass that raises the heat shielding effect by stacking multiple layers of silver is said to be popular. On the other hand, in the Tohoku and Hokkaido regions, the solar radiation acquisition type glass that takes in solar radiation moderately by using only a single layer of silver is said to be popular. In both cases, the values are the same level with regard to the heat transmittance rate for evaluating heat transferability, and there is no difference in the insulation effect.

In Japan that has a small land area and where houses are compact, windows are required to be about 100 mm thick in many cases. To realize adequate heat insulation even with this thinness, it is important to partition the hollow layer of the PairGlass into multiple layers to suppress air convection in the hollow layer. The triple glass shown in Fig. 3.10 uses three glasses and enhances heat insulation by enclosing argon gas or krypton gas in the hollow layer. Also, since the window consists of glass and frame (sash), the performance of the frame must also be improved. In future zero-energy housing, it is thought that windows that combine triple glass with thin high insulation resin sash and resin composite sash will be adopted.

Insulation properties of buildings have not only a significant impact on energy issues, but also on the health of their occupants. The Ministry of Land, Infrastructure, Transport, and Tourism conducted surveys in cooperation with the Ministry of Health, Labor, and Welfare and medical institutions on the death rate for each sea-

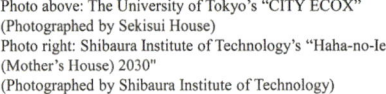
Photo above: The University of Tokyo's "CITY ECOX"
(Photographed by Sekisui House)
Photo right: Shibaura Institute of Technology's "Haha-no-Ie
(Mother's House) 2030"
(Photographed by Shibaura Institute of Technology)

Fig. 3.11 Case examples of triple glass
Model houses displayed at "Ene-Mane (Energy Management) House 2014," an exhibition themed on "houses in 2030" in which universities and companies collaborate to construct and display model houses. Both model houses incorporated low-e, gas filled triple glass manufactured by AGC Asahi Glass.

son by cause of death and the cause of accidents that occurred indoors in which injured persons were taken to hospitasl by ambulance. As a result, seasonal fluctuations were found in the cause of death due to the vasculature/cardiovascular system and the cause of death in residences. There are many cases of heat shock in the winter when the blood pressure rapidly changes due to temperature differences in the bathroom and dressing room. Houses with large temperature differences have a higher risk of people falling sick.

A report by Professor Shunji Ikaga of Keio University presented at the Architectural Institute of Japan is also very interesting. According to a questionnaire conducted for 5500 houses and 19,000 people living in highly insulated and highly airtight houses, there was a clear difference in the prevalence rate of illness before and after the people moved in. An 84% improvement was seen in cerebrovascular diseases, while an 81% improvement was seen in heart diseases. Improvements were also seen in all 10 diseases investigated including allergic rhinitis and atopic dermatitis.

Since insulated houses have uniform indoor temperature, vascular diseases such as heat shock are unlikely to occur. Also, since highly condensed and airtight houses are less susceptible to dew condensation that cause molds, it is thought that various symptoms such as allergy and atopy have improved. Higher insulation and higher airtightness of houses also lead to better quality of life (Figs. 3.11 and 3.12).

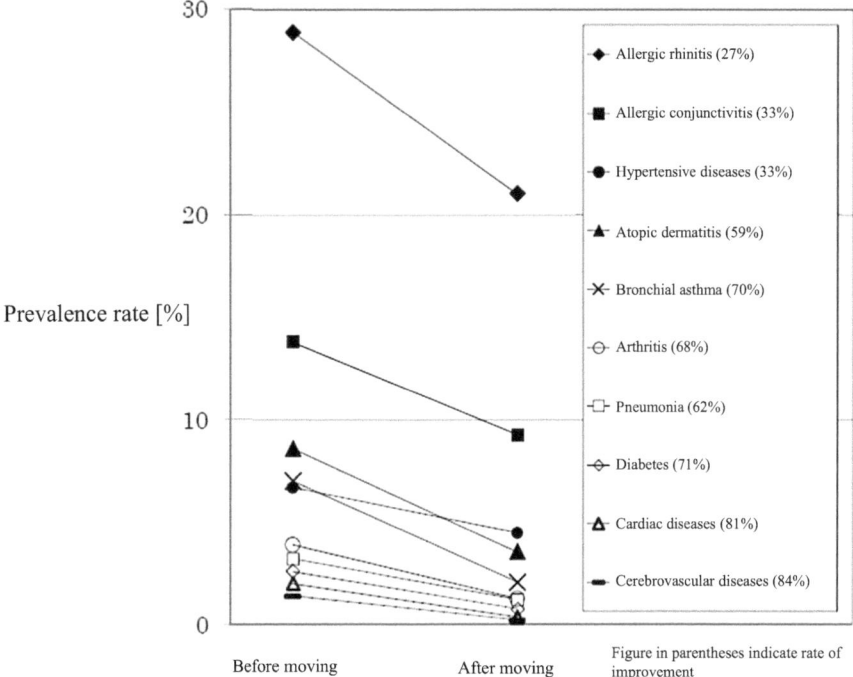

Fig. 3.12 Changes in disease prevalence rate due to improved insulation and airtightness performance and improvement rate. (Source: Ikaga et al. 2011)

3.3.5   The Latest Heat Pump Situation

Within the household, hot water supply, in particular, consumes a lot of energy. Japan is at the forefront of the world with regard to energy saving equipment in this field. One of these equipment is an electric water heater called EcoCute, which boils hot water with the heat of the air. EcoCute is based on a heat pump technology that enables extracting energy using temperature differences.

Currently, the power generation efficiency of thermal power plants in Japan is 42% on average. When you boil water with EcoCute, you can get about six times more energy, so multiplying 42% by 6 indicates that about 2.5 times more heat can be produced.

EcoCute has been on the market for more than 15 years, but during this time heat pump technology has evolved remarkably, prices have declined, and water heaters has consequently become increasingly popular. As shown in Fig. 3.13, the Heat Pump & Thermal Storage Technology Center of Japan has estimated the introduction of household heat pump water heaters in the future. The low variant represents

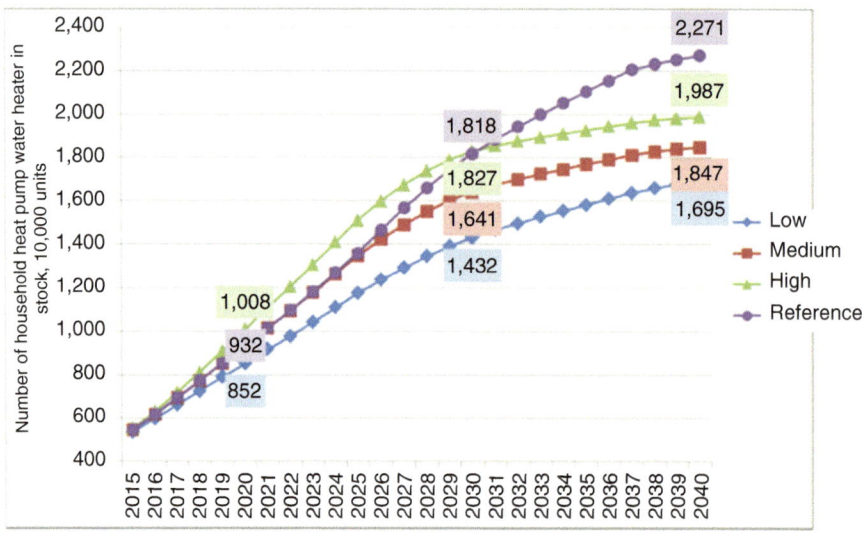

Fig. 3.13 Estimated number of household heat pump water heater introduced. (Source: "Survey on Prospects of HP Diffusion," Heat Pump & Thermal Storage Technology Center of Japan)

the current trend with no additional countermeasures, the medium variant represents the case where measures are taken such as subsidies for introducing the water heaters, and the high variant represents the case where stronger measures are implemented to accelerate their introduction. According to this estimate, there will be 3–4 times as much stock in 2040 compared to the current situation.

Heat pumps are used not only in EcoCute but also in various household appliances. Household air conditioners are a typical example. Due to technological progress over the last 20 years, the energy efficiency of air conditioners has doubled. Most freezer-refrigerators have a built-in heat pump, and heat pump type models are available for hot water floor heating, washing and drying machines, and hot water snow melting systems.

There are many heat pump type models in industrial air conditioning, freezer-refrigerators, and hot water supply equipment. Recently, cases of introduction of these equipments are increasing even at monozukuri sites. With the initial technology, it was possible to raise the temperatures of water to about 60 degrees (Celsius), so the industrial use was limited, but recent technology has made it possible to raise it to 90 degrees. If heating temperatures to this extent can be secured, these equipments can be used for heating and drying foods, as well as at chemical plants, electronic parts factories, and pharmaceutical factories (Fig. 3.14).

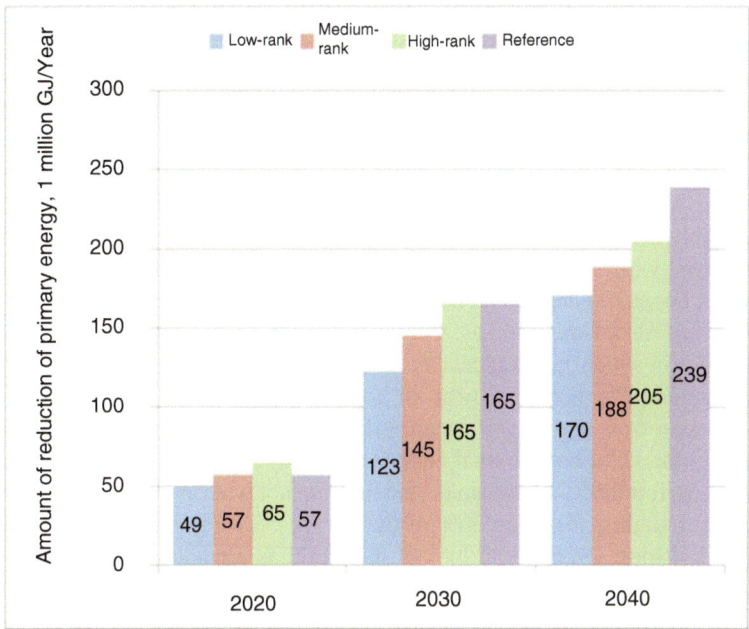

Fig. 3.14 Estimated energy-saving effects of introduction of household heat pump water heaters. (Source: "Survey on HP Diffusion Forecast," Heat Pump & Thermal Storage Technology Center of Japan)

3.3.6 *Domestic Fuel Cells Packed with Japanese Technologies*

Along with the electric water heater EcoCute, household fuel cells known as Ene-Farm are promising as energy saving equipment that can be used for hot water supply in the home. There are two types of fuel cells that supply electricity and heat: solid polymer type using hydrogen and solid oxide type directly using city gas. The first type of Ene-Farm refers to a product that extracts hydrogen from city gas, LP gas, kerosene, etc. and produces electricity when the hydrogen reacts with oxygen in the air. About 37% of the gas and oil input to produce hydrogen is converted into electricity, while 50% of the thermal energy generated in the process is used to produce hot water. Therefore, a total of 87% of the input energy can be used. Only 13% is lost.

Since the efficiency of electric power reaching the home from centralized thermal power plants is about 37%, the amount of electricity generated by domestic fuel cells has reached the level of centralized thermal power generation. Therefore, using the hot-water supply of domestic fuel cells is like using waste heat thrown away at a centralized thermal power plant.

EcoCute and Ene-Farm are a crystallization of the energy-saving technology in which Japan takes pride. EcoCute is a product that Japan made almost from zero.

Refrigerators and air conditioners once used CFC as a refrigerant. Refrigerant is indispensable for pumping in and pumping out heat, but CFC has been problematic, as it is an ozone depleting substance. Japan has succeeded in replacing it with CO_2.

Ene-Farm is more technically advanced. Technologies for extracting hydrogen from LP gas and ceramics technology, which is indispensable for fuel cells, can be considered an area where Japan has expertise, and their elemental technologies are refined day by day. Products that are close to power generation efficiency of 45% will probably come out soon.

Ene-Farm, which uses solid oxide fuel cells, generates electricity at a high temperature of about 700 degrees (Celsius), so power generation efficiency is high, and it is 52% in Osaka Gas products. Ceramics is used in this type of fuel cells, and technologies in which Japan excels are utilized.

EcoCute and Ene-Farm use heat for hot water supply. Hot water storage tanks made in Japan have higher performance than overseas products. There are similar products in the U.S., but the final energy efficiency is inferior due to low thermal insulation performances of the tanks, and they are not widespread. Hot water supply accounts for slightly less than 30% of the energy consumption of households and business divisions. Even if you look at the world as a whole, the demand for hot water for taking a shower or bath is quite large. This reduction in energy consumption greatly contributes to the improvement of energy efficiency and low carbonization.

3.3.7   Globalize Japanese Environmental Technologies

Japan, which leads the world in heat pump technologies in the fields of air conditioning and hot water supply, should contribute to the world's low carbonization with these advanced technologies, and at the same time, this should also lead to Japan's economic growth. Among Japanese companies, Daikin Industries, which is at the forefront of such technologies and boasts the world's largest share in the field of air conditioning, has placed global strategy as the pillar of its corporate management. The key points of the company's success can be summarized into the following three points.

1. **"Open Technology Strategy" – Strategy for Promotion of Energy-Saving Air Conditioners through Inverters -**

Daikin Industries developed an "Open Technology Strategy" for expanding the inverter air conditioner market by providing inverter technology, which is the essential part of the company's technology, to Gree Electric Appliances, the company with the largest share in China. As a result, in China's air-conditioner market, the ratio of inverters, which is about 30% more energy-saving compared to non-inverter machines, has rapidly expanded, and the wave of inverterization is spreading throughout Asia.

2. **Simultaneous development of refrigerant and air conditioners – Speedy response to environmental issues -**

Daikin Industries is the only manufacturer in the world to both manufacture and sell air conditioners and refrigerants. As global environmental regulations became stricter, the company quickly adopted refrigerant control that matched the nature of the refrigerant, thereby achieving both energy saving and environmental performance at the same time. As a result, Daikin developed a new refrigerant called "R32," which reduced the global warming potential to about one-third of existing refrigerants. The company developed the world's first room air conditioners that adopted this refrigerant, and they are sold in 43 countries as of March 2015.

3. **Split/multi-split type air conditioning – Meticulous refrigerant control technology -**

"Multi-technology" for separately controlling multiple indoor units with one outdoor unit to selectively use cooling and heating is realized by Daikin's refrigerant control technology. The company started development and sales of multi-split air conditioners for buildings in Japan in the 1980s, and realized the flexibility of construction and packaging of design and construction. Based on domestic success, the company has deployed its business model to Europe, Asia, and China, and spread Japanese style air conditioning culture.

In addition to Daikin Industries, other major air conditioning manufacturers in Japan such as Toshiba, Hitachi, and Mitsubishi Electric are rapidly advancing globalization through alliances with foreign companies.

Meanwhile, EcoCute, which succeeded in commercialization in the world for the first time in 2001, exceeded 5 million in the number of units sold at the end of March 2016, but its global expansion has been delayed.

Pressed to respond to environmental issues, South Korea and China are also aiming at switching from boilers heat pumps for hot water supply and heating. There is also concern that Japanese heat pump hot water supply equipment will become obsolete unless Japanese companies steer their global expansion as they did with air conditioning.

Under these circumstances, Panasonic began collaborative research with RWTH Aachen University in Germany in "electric power management technology for heat pump hot water heating systems." In addition, Mitsubishi Electric transferred the design and development functions of heat pump heaters marketed in Europe to a subsidiary in the UK, and aims at increasing market share by speeding up the introduction of new models and reducing costs. Such moves toward globalization have been emerging.

In the future, it is expected that heat pump hot water heaters will bloom at once in China, which is a huge market, and the global strategies of Japanese companies, including technology and sales alliances with local companies and foreign companies, will be further tested.

3.4 Low-Carbon Technologies for Monozukuri

3.4.1 Shift from Blast Furnaces to Electric Furnaces

While agriculture, forestry, and fisheries make use of biological resources derived from solar energy, monozukuri (making things) is an activity to produce various artificial objects using underground resources that did not exist in the material cycle of the natural world. Here, I would like to take up steel, which has the highest energy consumption in the monozukuri process.

Steel can be produced by the blast furnace process (hereinafter referred to as blast furnace process) or the electric furnace process. Iron ore, which is a natural resource, is used in the blast furnace process. Iron ore refers to iron oxide. If iron ore is cast into a blast furnace together with coke, which consists primarily of carbon, pig iron is produced as oxygen is removed from the iron oxide. Various steel products are created by processing this pig iron.

About 600 kg of coke (carbon) is used to produce one ton of iron. Theoretically, the required amount of carbon is 202 kg, so it means that two-thirds are not used. Although the overall iron and steel process has become more efficient because of technological development, the processing and molding process is a multi-stage process, so if a slight loss that occurs at each stage accumulates, this could turn into a significant loss. In the future, to what extent efficiency can be increased depends on investment, but it is not easy to reduce 600–400 kg.

Meanwhile, iron that is used in bridges, buildings, railroad tracks, automobiles, etc. that exist in society is recovered as scrap once its service life as a product comes to an end. The scrap iron, however, is melted in electric furnaces, molded, and used as steel products once again. The energy used in the electric furnace process to produce one ton of iron can be calculated in terms of the amount of carbon required for the process. This is about 300 kg, about half of the amount required in the blast furnace process.

Energy required for processing and molding cannot be reduced significantly even in the electric furnace process, but the direction of reducing the overall energy consumption is visible. Currently, electricity is used to melt the scrap iron. Fossil resources are burned, and the heat generated is converted into electricity, which, in turn, is converted into heat to melt the iron. If the iron can be directly melted with the fossil resources, the amount of carbon used per one ton of iron can be reduced to about 150 kg.

Thinking in terms of energy, rather than mining iron ore and producing iron in blast furnaces, far less energy is consumed by taking advantage of scrap iron and producing iron in electric furnaces. In addition, the direction of technology development to reduce the energy consumption is visible in the case of the electric furnace process. This applies not only to iron but also to cement, aluminum, etc. If a social system in which resources that lie under urban mines are recovered and recycled can be created, it is possible to significantly reduce carbon in the monozukuri sector.

3.4.2 Aluminum Is an Excellent Material in Terms of Recycling

Aluminum is produced from an ore called bauxite, which consists mainly of aluminum hydroxide (about 25% of the aluminum content). The energy used in creating the product can be broken down into the following process: Mining (transport), production of aluminum oxide from bauxite, reduction (electrolysis), and molding.

As described above, no energy is theoretically required for transport. The same hold true for molding and processing. If a material is heated to near the melting point, it becomes soft and can be easily molded. If heat is recovered when it cools, this heat is the same amount as that required during the heating process. If this recovered heat is used to heat the next sheet, and the process is repeated, no additional energy would be needed.

Cutting is among one of the processes, but if the material is melted and then solidified in separate portions, the only energy required would be the heating process, and if that heat is recovered, no additional energy would be required. There would be zero theoretical energy required for processing thick sheets into thin sheets, cutting, machining, molding or processing.

To produce one ton of aluminum, 20 GJ of energy would be required in the process of producing aluminum oxide from bauxite. In the relevant reduction (electrolysis) process, carbon electrodes that will wear out will be used, so together with the power required in the process, 130 GJ of energy will be used. This brings the total to 150 GJ. It is a high value that is five times the theoretical energy.

Recycling of aluminum is also very popular. According to the Japan Aluminum Association, the energy used to produce regenerated mass that is made by recycling scrap, etc. is about 3% of that required to produce new bullion from bauxite. Using scraps is far more efficient than using ores, and it can be said that the scraps are an excellent resource in terms of energy consumption.

3.4.3 Achieving Material Cycling of Rare Metals

The existence of urban mines is extremely important in low carbonization in the monozukuri industry. Urban mine is a concept that was proposed more than 30 years ago, but it came to be recognized by the general public only about 10 years ago when the prices of rare metals that are indispensable to electronic devices such as mobile phones and personal computers began soaring.

Rare metals are literally very scarce, and there are many mining areas that are unevenly distributed. For example, small amounts of dysprosium are added to neodymium magnets used in hybrid and electric vehicle motors in to prevent deterioration of their magnetic force. Nearly all the dysprosium on the earth is in China, and if imports stop, it will be no longer be possible to produce hybrid vehicles and electric vehicles. Urban mines are important not only in terms of low carbonization but also in terms of resource security.

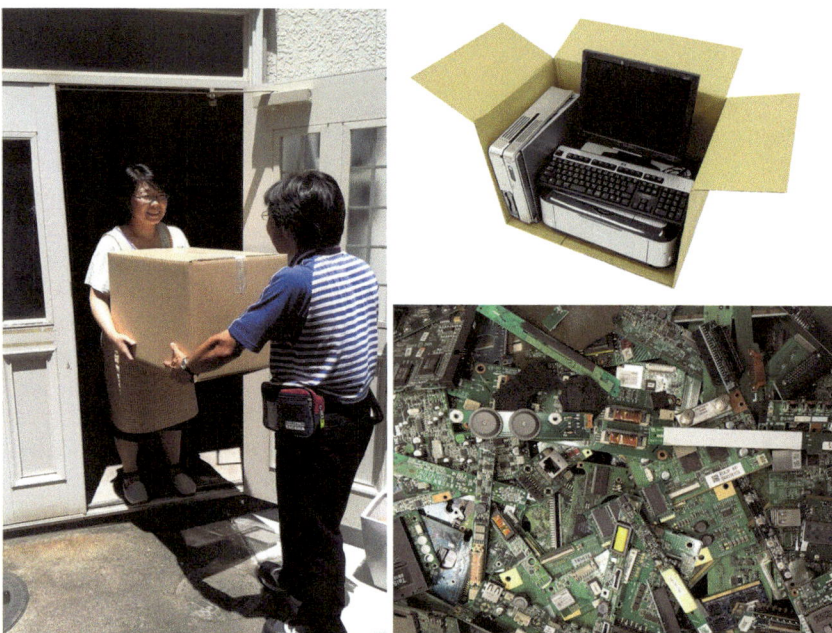

1) Apply via the Internet in advance and have a home delivery service pick up appliances
2) Courier fee is ¥880 for one cardboard box (excluding tax). Any number of small household appliances may be put in the box.
3) The small household appliances that have been collected are properly taken care of by an intermediate treatment company
   (certified service provider) that has a business tie-up with Lynette Japan; the appliances are taken apart, separated into different
   parts, and recycled.

Fig. 3.15 Small household appliances collection service using home delivery services. (Courtesy: Lynette Japan Co., Ltd.)

Fortunately for Japan, where electronic devices are widespread, large amounts of rare metals are dormant, so it is said that if all unused products can be recovered, these urban mines alone can satisfy domestic demand. To that end, it is of course necessary to establish a mechanism for material cycling.

It was precisely for this reason that the Small Home Appliance Recycling Act was enacted in April 2013. In the past, small appliances such as personal computers, mobile phones, smart phones, digital cameras, and gaming devices were not subject to recycling laws, and had been classified as non-combustible waste. The intention here is to turn these appliances into resources by properly creating a mechanism of recycling.

Renet Japan, a certified operator under the Small Home Appliance Recycling Act, offers the service of collecting small home appliances such as personal computers and smartphones and recycling them by taking advantage of couriers and the Internet (Fig. 3.15). The key feature of this service is the convenience that consumers find in being able to apply for the service from their home any time around the clock and having courier companies come to their home for pick up, as well as data security, or in other words, the thorough implementation of complete erasure of stored information, so that consumers can request for the service with confidence.

Renet Japan has established partnerships with about 100 municipal governments in 2 years after it started the service. These municipal governments notify their residents about the company's courier and collection service as part of their administrative services. In Kyoto City, about 10,000 personal computers were collected in just 2 months.

Since the Small Home Appliance Recycling Act is a law aimed at promoting recycling and lacks penal regulations, there are reportedly some areas where recovery of small home appliances has not progressed sufficiently. Although this is an administrative service, the cooperation of the residents cannot be gained if it is too troublesome to use the service, or there remain concerns about the handling of data. It is precisely in this area that ideas that are unique to the private sector regarding such services should be taken advantage of.

3.4.4   Expectations for Dissemination of Industrial Heat Pumps

At monozukuri sites, ordinary boilers are generally used in processes where heating is required. Among them, such applications as hot water supply, washing, drying, and low temperature heating (fermentation, aging, etc.) require heating below 100 °C. In such cases, it may be possible to replace the boilers with heat pumps. In the past, large-scale centralized equipment such as large boilers were used, but by replacing them with distributed-type equipment such as industrial heat pumps, it may be possible to achieve significant energy savings by reducing waste such as piping loss and drainage loss.

Figure 3.16 shows an example of heat pump application. Conventional heat pumps could be used for heating up to about 60 °C, but now that high-temperature heat pumps have been developed, their use in the food sector has spread remarkably. In addition, processes in which they can be used have increased in the medical, chemical, and other fields.

Industrial heat pumps can be roughly classified into two types depending on the positioning of the heat to be used.

One of them is the waste heat recovery type heat pump that effectively utilizes the heat that used to be discarded. At monozukuri sites, waste heat is recovered and utilized in various situations, but in some cases, the heat is irrecoverable and released into the air. The heat pump can effectively use waste heat as a heat source for temperature zones around several tens of degrees that are difficult to use. In the case where there is a time lag between the time when heat is generated and the time when heat is used, thermal energy can be utilized without waste by installing a thermal storage tank for storing the generated heat.

The other type is a heat pump that performs simultaneous heating and cooling. Although there are not a few cases at food factories, etc. where both the cooling and the heating steps are present, the heat pump is originally a technology in which heat is removed from one side and emitted to the other side. The side from where heat

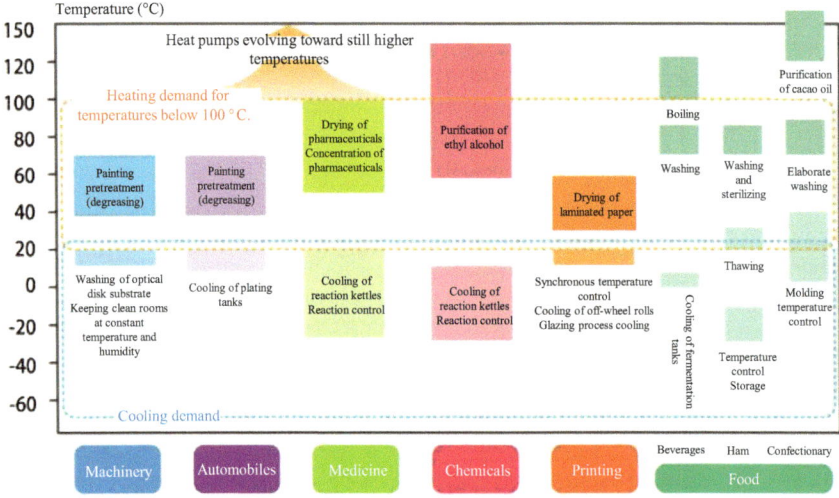

Fig. 3.16 Heat pump application destinations. (Source: Heat Pump & Thermal Storage Technology Center of Japan)

has been pumped out is cooled, while the other side where the heat has been pumped in is heated. Energy efficiency is nearly doubled if cooling and heating are performed simultaneously instead of separately. This would lead to energy saving within an entire plant.

For example, at a noodle factory, water is boiled in a boiler when noodles are boiled. The boiled noodles are cooled in a chiller filled with cold water. By installing a heat pump between these two steps, hot water can be produced on one side by collecting heat, while cold water can be produced on the other side by removing the heat. A certain factory succeeded in reducing CO_2 emissions by 31% and energy consumption by 35%.

In 2015, Japan drew up draft pledge aimed at reducing greenhouse gas emissions beyond 2020. The target is to achieve a 26% reduction in greenhouse gas emission levels (about 1.042 billion tons of CO_2) in FY2030 compared to FY2013 (approximately 25.4% reduction compared to FY2005). The Heat Pump & Thermal Storage Technology Center of Japan predicts dissemination of heat pumps for domestic and industrial use. It has estimated the effects of energy savings in the medium-level case and the impact on the reduction target in draft pledge (Fig. 3.17). According to the estimate, more than 20% of the target can be achieved by increasing introduction of the heat pumps.

Theoretically, it has been known that a heat pump that pumps up heat from the environment and utilizes that energy is correct. That is because it is inefficient to burn fossil fuels to secure low temperatures, such as a hot water supply. Places that directly burn resources will likely disappear more and more in the future.

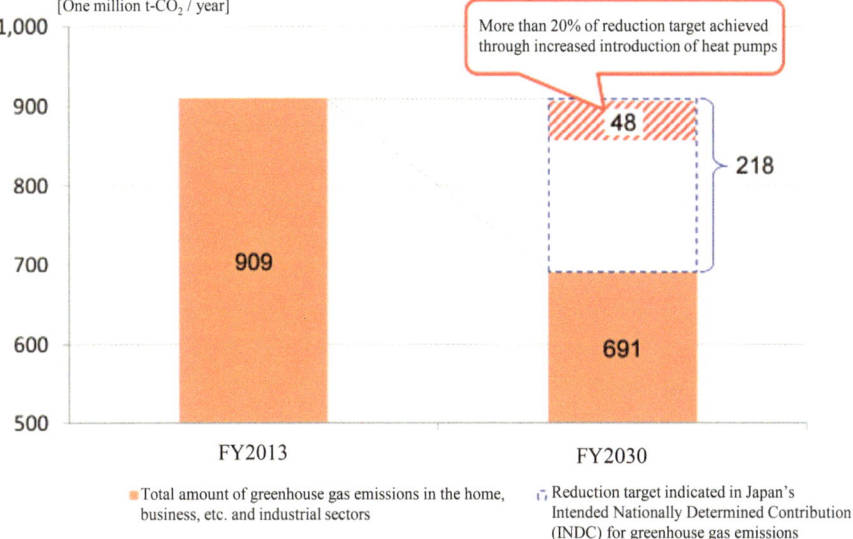

Fig. 3.17 Effects of CO_2 reductions through the introduction of heat pumps. (Source: Heat Pump & Thermal Storage Technology Center of Japan)

Bibliography

Agency for Natural Resources and Energy, Japan (2017) FY2017 Annual report on energy (Energy white paper) (JP only)

Heat Pump & Thermal Storage Technology Center of Japan (2015) Survey on prospects of HP diffusion

Ikaga T, Eguchi R, Murakami S, Iwamae A, Hoshi T et al (2011) Evaluation of investment in residential thermal insulation considering non-energy benefits (NEB) delivered by health. J Environ Eng (Transactions of AIJ) 76(666):735–740

Ministry of Land, Infrastructure, Transport, and Tourism, Japan. http://www.mlit.go.jp/sogosei-saku/environment/sosei_environment_tk_000007.html. Accessed Jan 2018

Chapter 4
Technology to Support Low-Carbon Society (Utilizing Energy)

4.1 Future Image of Renewable Energy

4.1.1 Rethinking the Value of Renewable Energy

In the previous chapters, we have discussed technologies that increase energy efficiency, while minimizing energy consumption and maximizing the obtainable benefits. However, regardless of how much effort is made to improve efficiency, energy consumption does not become zero, so energy resources with low environmental impact must somehow be secured.

Firstly, energy resources represent an original source of energy that you do not have to get energy from anything else. They must be available from nature, such as buried underground, growing on land, or fallen from the sky.

According to this definition, fossil resources, sunlight, hydropower, wind power, geothermal power, and biomass are all energy resources. However, hydrogen and electricity are not energy resources, since hydrogen and electricity cannot be obtained from nature.

However, hydrogen is often misunderstood. It has been said that "hydrogen will solve the energy problem"; "hydrogen is obtained by electrolyzing inexhaustible water, so if we can obtain energy with hydrogen, energy problems will be solved"; and there has been discussion of a "Hydrogen nation". Certainly, water is inexhaustible and hydrogen can also be used for power generation. Hydrogen can be used for both thermal power generation and fuel cells, and it can be regarded as a resource because it becomes the source of electricity. However, to make hydrogen, we need another resource. Therefore, hydrogen cannot be considered an energy resource of itself.

Looking at the amount of energy resources used, the fossil resource usage rate is still so overwhelmingly large that non-fossil resources account for 24% worldwide and 7% in Japan. However, in order to realize a low-carbon society while maintaining the economy for mankind, we must increase the proportion of non-fossil

© The Author(s) 2018
H. Komiyama, K. Yamada, *New Vision 2050*, Science for Sustainable Societies,
https://doi.org/10.1007/978-4-431-56623-6_4

resources. To that end, it is necessary to expand the use of renewable energies such as sunlight, wind power, and biomass.

As mentioned in Chap. 2, considering the progress since 1995, "doubling renewable energy" as proposed in Vision 2050 should be revised upward to "half the total energy should come from renewable sources." Let's look at specific strategies to achieve this goal.

4.1.2   The Future Image of Solar Cells and Storage Batteries

In the long run, the main power supply will become renewable energy rather than fossil fuels. There are many types of renewable energy such as hydropower, wind power, geothermal power, and biomass. Among these, solar power has enormous energy availability, and there is also room for cost reduction through future technological development. Thus, expectations are growing for expansion of usage.

However, the amount of power generation varies under the natural cycles of sunlight. In summer and winter, the sunshine hours and sunshine intensity are different, and the weather changes from day to day. In addition, although power generation is limited to daytime but usage takes place in daytime and nighttime and there is thus a gap between the electricity supply and the timing of demand. In order to compensate for this, it is necessary to use a storage battery in combination with the photovoltaic technology. Therefore, not only the photovoltaic power generation system but also the future trends of the storage battery must be clarified in order to study and design the future power supply configuration.

There are various research and reports on cost analysis such as past trend analysis, prospects and scenarios announced by associations and administration, technical roadmaps, market research, and economic analysis. However, these are mainly evaluations of economic efficiency based on experience curves (learning curves). Changes in raw materials, processes, and production scales due to future technological development are not sufficiently discussed, hence it is impossible to project the manufacturing cost.

The method that we applied to make the projections described below puts emphasis on clarifying concrete technical contents to calculate current and future costs. Then we will design the manufacturing process including a detailed equipment list and quantitatively evaluate the economics and environment of products and systems based on the results.

Also, since the influence of the cost of the production scale and the technology level is large, the relationship between these factors will also be clarified. The speed of future technical advancement will be predicted from the progress of related technologies.

This evaluation approach is also useful for planning the investment time and production scale in the product manufacturing plant.

4.1.3   Importance of Balance Between Future Cost and Investment

First of all, let's look at the evaluation result for solar power generation systems.

Table 4.1 shows the prediction of the performances and costs in 2020 and 2030, referring to 2012 as the base year. The current mainstream, single crystal silicon solar cell (single crystal Si), has a module conversion efficiency of 17%, whereas 15% can be achieved with a compound thin film solar cell (CIGS), which has an expanding market. However, it is expected that by 2020, CIGS technology will improve and power generation efficiency will rise to 18%, and annual production volume will increase, while manufacturing cost could be reduced. Furthermore, apart from CIGS, a new type of thin-film solar cell will also be introduced, and conversion efficiency of this is predicted to be about 15%.

The thickness of monocrystalline Si would be suppressed to less than 1/3 of the current level, and power generation efficiency will increase to 25% by 2030. In addition, the tandem CIGS type with the layered structure of multiple CIGSs will expand the market with the high performance of 30% conversion efficiency as a major strength.

Table 4.1 Breakdown of solar power generation system costs

Technological level		Current situation in 2015		2020		2030	
Solar cell		Single-crystal Si 150 µm thickness	CIGS	CIGS	New thin-film	Single-crystal Si 50 µm thickness	New CIGS tandem
Module conversion efficiency		20%	15%	18%	15%	25%	30%
Amount of annual production (GW/Year)		1	1	5	1	5	5
Production cost	Variable costs (raw material costs)	56	51	40	34	35	29
	Variable costs (utilities expenses)	4	2	1	2	1	1
	Fixed costs (equipment costs and personnel expenses)	14	14	9	12	6	7
	Module subtotal (Yen/W)	74	67	50	48	42	37
BOS	Frame (including construction costs)	22	29	27	32	12	10
	Power conditioner	30	30	20	20	10	10
	BOS subtotal (Yen/W)	52	59	47	52	22	20
System (Yen/W)		126	126	97	100	64	57

Source: Created based on materials from the Center for Low-Carbon Society Strategy, Japan Science and Technology Agency

The manufacturing cost of the whole system is expressed by the sum of the manufacturing cost of the module and the balance of system (including peripheral equipment and construction). The manufacturing cost per 1 watt (W) in 2015 was ¥126. It is predicted that it will fall to ¥97–¥100 in 2020, and to ¥57–¥64 in 2030.

Secondly, consider the power generation cost. The production cost of single crystal Si in 2015 was ¥126/W. Assuming that the annual expense rate required to operate this system is 10% of the manufacturing cost, the annual expenses will be ¥12.6/W. Since the annual power generation amount of the 1-W system is 1 W × 1000 hours (h) = 1000 Wh (1 kWh), if the annual sunshine amount is 1000 h, the power generation cost is ¥12.6 per 1 kWh.

Various values have been reported as the power generation cost worldwide, and some are lower than ¥12.6/kWh. For example, there are many reported values such as ¥3 to ¥5/kWh in Dubai and the U.S. where there are good sunshine conditions. In these areas, annual sunshine amounts are large and the amount of electricity generated is more than twice as large as those generated in Japan. Therefore, if the annual expense ratio is set to 7% of the manufacturing cost, the generation cost will be less than ¥5/kWh. Our cost calculation results are in line with the reports from other locations worldwide.

Figure 4.1 shows trends in the cost of solar power generation module and system production along with the sales prices of modules in Japan and China. We produced

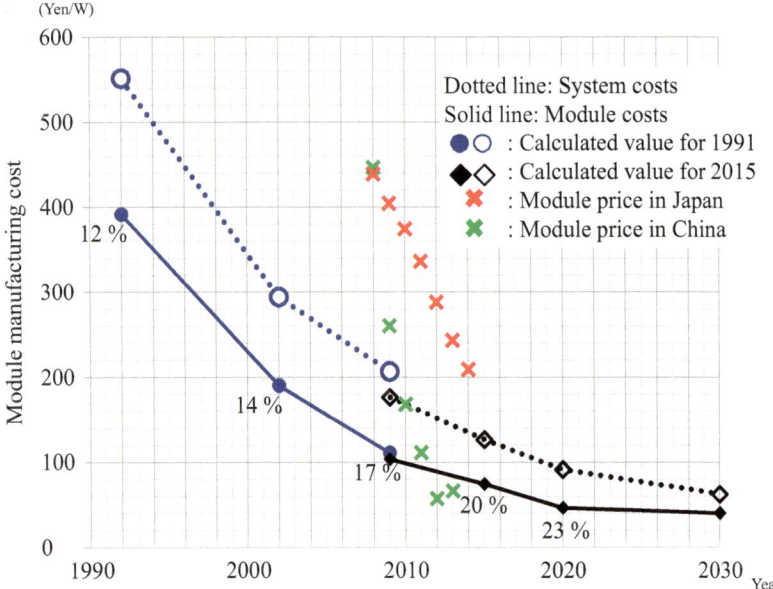

Fig. 4.1 Manufacturing costs of Solar power generation module system. (Source: Created based on materials from the Center for Low-Carbon Society Strategy, Japan Science and Technology Agency)

the cost calculations of the future products in 1991, and values after 2010 were calculated by the Low Carbon Society Strategy Center. From 2010 onwards, due to market expansion and intensified competition, the module price sharply declined and it is now approaching the cost calculation value. In particular, despite the effects of foreign exchange and national policy, the selling price of the modules made in China is lower than the cost. Following the U.S., an anti-dumping and anti-subsidization tax was imposed in Europe in 2012 on solar panels made in China, but it seems there is a logical basis for this based on our cost calculation.

It is quite severe for companies if the selling price drops by this amount. Table 4.2 shows the current profit status of foreign solar cell manufacturing companies. Most companies have fallen into deficit.

From the above discussion, we can see the important considerations needed when companies make investments. As technology advances, the market expands and the production volume increases, while the manufacturing cost decreases. However, this also applies to competitors. If price competition intensifies and the selling price falls below cost, only the deficit will keep accumulating.

That is why it is important to make management decisions by accurately projecting future technological progress, production scale, and future costs. It is meaningless to create a business plan by combining the future market size and the current manufacturing cost. In order to properly determine the technology selection, scale, and timing of a new factory, it is important to determine the future cost. Furthermore, when taking foreign policy such as anti-dumping taxation into consideration, such cost evaluation clearly becomes indispensable.

Table 4.2 Profits of the solar cell module manufacturers

Company	Module shipment volume[a] in 2015	Net interest expenses in 2015[b]	Impact[c] of net interest expenses (net interest expenses ÷ shipment volume)	Remarks
	GW	M$	¢/W-module	
SunPower	1.3	**−42**	**−3.2**	US, n-type
JA Solar	3.1	**−39**	**−1.3**	China
Jinko	3.5	**−60**	**−1.7**	China
Trina Solar	4.5	**−49**	**−1.1**	China
Canadian Solar	4.1	**−37**	**−0.9**	Canada
Yingli	3.6	**−155**	**−4.3**	China defaulted
First Solar	2.5	16	0.6	US

Note: Leading-edge factory production cost: About 60¢/W
Source: Created based on materials from the Center for Low-Carbon Society Strategy, Japan Science and Technology Agency
[a]Shipment volume: Fuji Research Institute, [b]Materials presented by First Solar @PVSEC 2016. (Sources are each company's Annual Report)
[c]Since each company also sells cells alone, the figures per watt are reference values

4.1.4 Which Storage Battery Will Be Playing the Leading Role in 2050?

Lithium-ion batteries are used in personal computers and smartphones, and they can be regarded as one of the most familiar storage batteries in our everyday life. The basic structure of a lithium ion battery is a positive electrode composed mainly of lithium metal oxide and a negative electrode typically composed of graphite, which are immersed in an electrolytic solution together with a separator dividing both electrodes. When electricity flows, lithium ions from the positive electrode dissolve into the electrolytic solution, then run through the separator and move to the negative electrode side and remain between the graphite layers. This is a state of electrical charge. When discharging, lithium ions dissolve out from the graphite layer and move to the positive electrode side.

High stability and an economy that can withstand mass introduction are required for the combination of fixed storage batteries with solar power generation. Lithium ion batteries have higher energy density, longer life, and charge–discharge efficiency as high as 90% compared with conventional storage batteries such as lead storage batteries and nickel hydride batteries. It can be said that this is an easy-to-introduce storage battery, as material development and manufacturing technology have improved and prices have also declined in current expanding period.

The prices of current household storage batteries are shown in Table 4.3. The selling price of a Japanese company is about 10 times the manufacturing cost of ¥13.9/Wh shown in Table 4.4, while the selling price of Tesla in the U.S. is about three times this value. Severe price competition will start as the market develops, due to such factors as expansion of production scale, improvement of product yield, and development of new products.

In contrast to lithium ion batteries that have already become widespread, NAS batteries are expected to become popular in the future. This battery was named "NAS" because sodium (Na) is used for the negative electrode and sulfur (S) is used for the positive electrode. A fine ceramic solid electrolyte membrane is used between

Table 4.3 Manufacturer and price of household energy storage systems(Li ion battery) (as of May 2015)

Manufacturer	Product type number	Capacity[kWh]	Price [10,000 Yen]	Unit price [Yen/Wh]	Real price (subsidies included) [Yen/Wh]
Sharp	JH-WBP07A	4.4	107	240	150
Panasonic	PLJ-25522K	5.6	119	210	135
Eliiy Power	EPS-1 1	6.2	131	210	130
NEC	ESS-H-002006B2A	5.53	123	220	140
Tesla	Powerwall	10	41	41	–

LCS cost calculation value of Tesla products: ¥15/Wh (Battery: ¥13, BOS: ¥2)
Source: Created based on materials from the Center for Low-Carbon Society Strategy, Japan Science and Technology Agency

4.1 Future Image of Renewable Energy 69

Table 4.4 Current situation of lithium-ion batteries and future scenarios

			Current situation	2020	2030
Battery type			LiNi-C based	LiNi-C based	Li-S based
Production scale [GWh/y]			1(10)	10	10
Yield [%]			66(90)	90	90
Energy density [Wh/kg]			250	340	530
Active substances (positive electrode/negative electrode)			$LiNi_{0.85}Co_{0.12}Al_{0.03}O_2$/graphite	$LiNi_{0.85}Co_{0.12}Al_{0.03}O_2$/graphite	S-C/Li
Positive/negative electrode capacity density [mAh/g]			200/300	270/380	1500/2900
Comparison of actual capacity of positive/negative electrodes and theoretical values			0.71/0.78	0.97/0.99	0.9/0.75
Production costs [Yen/Wh]	Variable costs	Raw material costs	10.2(7.5)	4.8	5.5
		Utilities expenses	0.5(0.4)	0.4	0.1
	Fixed costs		3.2(1.7)	1.4	1.0
	Total		13.9(9.6)	6.6	6.6

Source: Created based on materials from the Center for Low-Carbon Society Strategy, Japan Science and Technology Agency

the positive and negative electrodes. The battery reaction occurs when sodium ions dissolved from the negative electrode combine with sulfur at the positive electrode to produce sodium polysulfide. At the time of charging, the bonds in the sodium polysulfide are broken, and sodium ions move to the positive electrode. The charge–discharge efficiency is about 80%.

The typical characteristic of NAS batteries is that they are high capacity batteries with high energy density. If capacity is the same, the size of an NAS battery can be only about one third the size of a lead storage battery. An NAS battery can handle megawatt-grade electricity storage and it is suitable for large scale solar power systems when arranged in large quantities.

Another potentially appropriate battery is the redox flow battery. This battery contains a tank containing an electrolytic solution with a potential difference between the positive electrode and the negative electrode. There is an electrolysis cell between the two poles, which are connected by a pipe and pump. Historically, there are various active materials used for electrodes, but currently the most promising type uses vanadium for both the positive and negative electrodes.

While other storage batteries exchange different ions at the electrodes to charge and discharge, the redox flow battery charges and discharges by the oxidation reduction reaction of the electrolytic solution, so the battery capacity hardly drops. In other words, long-term use is possible while maintaining the performance. However, the charge–discharge efficiency is only 75% and the energy density is not high. Therefore, although it is not suitable for miniaturization, it has the major advantage that there is no risk of thermal runaway and ignition.

4.1.5   Promising Markets Where Various Uses Can Be Considered

We introduced three types of storage batteries, but here we will describe the future prospects for lithium-based storage batteries with the highest charge–discharge efficiency. Table 4.4 shows the current state and future scenario of lithium-ion batteries. Currently, nickel-based (Ni type) batteries with energy density per kilogram of 250 Wh can be manufactured at ¥13.9/Wh. It is forecast that technology will advance by 2020, and energy density will be increased to 340 Wh/kg, while manufacturing cost will be reduced to ¥6.6/Wh. In 2030, the active materials will switch to the Li-S system, the energy density will increase to 530 Wh/kg, but the cost cannot decrease substantially.

Unlike conventional lithium ion batteries, lithium-air batteries do not use metal compounds, but react with metal lithium and oxygen in the air to generate electricity. Because of the high energy density of these batteries, the automobile industry is paying attention to this upcoming technology as it can be made smaller and lighter. Nissan's EV "Leaf" is equipped with a lithium-ion battery weighing about 300 kg

Fig. 4.2 Small hydroelectric power generation using agricultural water. (Courtesy: Imagineer Co., Ltd.)

or 20% of the car's weight, which is inefficient in terms of weight and cost. An EV requires a battery with three times the current energy density. As shown in Table 4.4, it will be closer to this target in 2030, and may be beyond it in 2050. In 2050, if 10 million EVs are running in Japan, there will be a large storage battery market of about 500 GWh just for the EVs.

Meanwhile, the requirement for the physical properties of household storage batteries and power storage batteries is not as stringent as that for automobiles. The size of the storage battery market for power supply is about the same as that for EVs. Furthermore, many other applications such as for use in robots can be considered, so it can be said that the battery market is a particularly promising field.

As such, I have described the quantitative calculation results including the current state and future prospects of solar cells and storage batteries. Economy is necessary for expanding the market, and for that purpose, high efficiency and high performance are indispensable. The research and development issues are clear, and a large number of researchers and engineers are working on this issue worldwide. Although competition between companies will intensify, it is a market that will continue to grow for decades (Fig. 4.2).

I now introduce an interesting corporate strategy for commodities. The Dow Chemical Company, a huge petrochemical company, has been successful in the commodities field. The outline of its strategy is as follows:

1. Keep the top share, or continue securing the share within the top three;
2. Maintain superiority, even if differences with other companies in terms of quality and manufacturing cost are actually small;
3. To achieve the above, continue investing no matter how bad the economic environment is;
4. On top of that, it is important to make a society that is not afraid of failure.

In this way, Dow started to dominate the commodity field. In the past, both solar cells and storage batteries may have been classified as high-performance products, but in the future, they will be commodities. Technological development is indispensable, but a management strategy like that of Dow is also essential. We expect that some Japanese companies will succeed in this field. Furthermore, We hope that new technologies to enhance energy efficiency of new product and to reduce manufacturing costs of them beyond what we forecast in this book will emerge.

4.1.6 Dissemination of Hydropower Generation by Region

Although Japan is said to be a country with limited resources, water resources are abundant enough to allow self-sufficiency. Hydropower is renewable energy that does not use fossil fuels, but it is difficult to build any new hydropower facility that is currently accompanied by large-scale development that is typical in Japan.

However, small hydropower generation has considerable potential for development. Even hydropower plants with an output of 10,000 kW or less are estimated to have a potential of about 10 million kW in the whole country. This is equivalent to about 4% of the total electricity generation capacity of 243.6 million kW (2010) of the general electric utilities, about 80% of the amount of power generated by large scale hydroelectric power plants, or 10 nuclear power plants.

Small hydropower plants have a propeller installed where there is a difference in water level and generate electricity using the thrust generated by the propeller. Places in which electricity can be generated include water supply and drainage pipes, rivers, and agricultural canals where water flows beyond a certain flow rate. Small hydropower generation has not become thoroughly widespread because of the high cost, even though the potential has been pointed out in the past. The ability to generate power in various places is an advantage of small hydropower generation, but it is also a disadvantage at the same time. Since conditions such as water flow rate, flow path, and the surrounding environment change according to the location, it is necessary to amend the dimensions of the propeller to suit each case. It would not be profitable if you were to build various types of equipment for a hydropower plant that generates about 10 kW. Conversely, it could be profitable to position a general-purpose propeller in a place where the installation environment is similar.

There are already moves to increase the number and capacity of hydropower projects in Japan. Imagineer Co., Ltd., known for the development of game content, is working on a plan to expand small hydropower generation to agricultural canals in collaboration with a land improvement district. According to Mr. Takayuki Kamikura, CEO of the company, demonstration experiments are being conducted in Miyagi Prefecture, and introduction of small hydropower plants with power generation capacity of 500 to 1000 kW are planned in Toyama Prefecture.

Since agricultural canals are generally similar in shape, if three or four propellers are made, they can be installed in many places. However, there are some technical hurdles, for example, the AC power generated by the propeller must be converted to DC, re-converted to AC, and then linked to the grid. Collaborative research between Yaskawa Electric Corporation and Japanese universities is being conducted to improve efficiency.

To disseminate small hydropower generation, we must also overcome the hurdles of institutions and regulations. Agricultural canals were originally built for agricultural use, and there are many water rights stakeholders. If these agricultural canals were to be used for power generation, some people may disagree saying that their agricultural applications would be affected.

To overcome these issues, Mr. Kamikura's project is taking place in collaboration with a land improvement district. His company works to maintain a relationship with the local people, municipal governments, and agricultural affiliated organizations, so it is easy to gain the understanding of the local community. Moreover, the maintenance and management of the power generation equipment is entrusted to the land improvement district. Power generation facilities are normally managed wirelessly, but staff members of the land improvement district will rush to the scene if an error occurs.

Small hydroelectric power generation can be realized through improvement of technology and the system, and can contribute to local production and consumption of energy. Even if the steps are small, accumulating achievements will lead to low carbon society.

4.1.7 The Potential of Biomass

Biomass has the potential to enable Japan to achieve a high energy self-sufficiency rate as in the case of water resources.

In 1999, when I proposed Vision 2050, biomass plantations were drawing attention in the world. However, in recent years, biomass has rarely been addressed in the public sphere. The cause of this decline in interest in biomass is economic rationality. Wood for construction is worth ¥50,000 per cubic meter, whereas wood for fuel is worth about ¥5000 per cubic meter, which is the same level as coal. That is about one tenth. Even if you were to operate a large-scale farm for biomass only, it would not be economically viable.

Fig. 4.3 Large-scale forestry in Sweden
A harvester in the process of bucking trees; it can perform felling and bucking of a single tree in about 40 s.

Therefore, when I proposed Vision 2050, I thought that we cannot expect much from biomass, but recently new possibilities have emerged. This is because many people have come to understand that expanding domestically produced wood consumption leads to healthy forests.

Currently, Japanese forestry has weakened as a result of losing in price competition with foreign countries. In Japan, 55% of timber products are used for construction and 40% for pulp and chips, among which only 30% of domestic resources are used. This amount represents only around 20–30% of Japanese forest resources. It is an important task to keep Japanese forest healthy and to reform forestry into a form that is economically feasible even with a self-sufficiency rate of 100% (Fig. 4.3).

Regarding forestry as a whole, we can learn from success stories such as supply chain building in Sweden or Austria. In the Platinum Society Network, a Smart Forestry Working Group has been established to discuss forestry revitalization, such as building a supply chain using wood. Forestry is an extremely important industry from the standpoint of building national resilience and existing in harmony with nature, as described in Chap. 6. It is not easy to regenerate a completely weakened industry, the whole country should tackle the problem of forestry revitalization as an approach to simultaneously solving both resource and environmental problems.

4.1.8 Hydrogen as a Partner of Renewable Energy

Renewable energy sources such as sunlight, hydropower, and biomass are mostly converted into electricity and used for daily living and monozukuri. Although electricity is not an energy resource, it is a form of energy that is easy for human beings to use and will become increasingly important in the future.

Hydrogen is not an energy resource just as electricity, but it can be used as a CO_2-free energy medium. For example, picture the following image: in combination with a power generation method that has fluctuations in the amount of power generated such as solar power, hydrogen is produced by electrolysis when the amount of power generated is large (i.e., in bright conditions) then electrical energy is stored in the form of hydrogen like a storage battery.

Gas cylinders containing hydrogen are not as heavy as storage batteries, and can be transported and stored. Therefore, it does not matter if the place where electricity is generated (the place where hydrogen is produced) and the place where it is used are separated. Hydrogen can be used directly as a CO_2-free energy medium for fuel cells and thermal power generation. Attempts have recently been made to reform lignite, which is an unused fossil fuel resource, into hydrogen. Lignite is carbonized plants from tens of millions to 100 million years ago. Because coal is formed from plants from 300 million years ago, lignite can be considered as young coal. However, lignite is not as user-friendly as coal, it is heavy and has high moisture content, so it is not suitable for shipping because of the high cost. Moreover, its physical properties are unstable, and it will ignite spontaneously when dried (Fig. 4.4).

Therefore, a method has been devised in which hydrogen can be extracted from lignite using a chemical reaction (gasification process) near the mining site. Since gas containing CO_2 is generated in this process, it is collected and combined with CO_2 Capture and Storage (CCS) for underground storage. Kawasaki Heavy Industries is promoting a project to transport lignite-derived hydrogen from Latrobe Valley, Victoria, Australia. There is a withering gas field in the vicinity of the mining site, which can be used for CCS. The concept is that hydrogen can be obtained from low grade coal without CO_2 emissions.

At the same time as this project, Kawasaki Heavy Industries is also working on the development of hydrogen-dedicated gas turbine thermal power generation facilities. Since only water is generated by burning hydrogen, zero emissions can be practically achieved through a series of supply chains, beginning with lignite.

Fig. 4.4 Lignite mine in Australia and lignite, which is a young coal. (Courtesy: Kawasaki Heavy Industries)

4.2   Innovations Emerging from Theory and IT

4.2.1   Pursuing Efficiency to the Utmost Limit

Overall, we are moving in the direction indicated by the three goals in Vision 2050. Innovations are necessary to achieve the three goals of tripling the energy efficiency, doubling the share of renewable energy, and establishing a material circulation system. However, innovation cannot occur by just waiting. As both scientists and engineers conduct research seriously and engage in technological development, they often say, "We cannot increase efficiency any further," or "We are doing our best." However, pioneers have opened up a new world by developing ways to overcome that wall.

One good example is pollution control in Japan and Germany. Due to the spread of pollution accompanying economic development since the 1950s, strict environmental regulations have been laid in Japan and Germany. Some believed that the regulations would cause a decline in the productivity of factories and would not be economically feasible. It certainly must have been difficult to overcome the situation by relying on the existing technology at the time.

However, as a result of working on research and development by back-casting toward the vision of halting pollution, a number of innovative ideas were born. Thanks to those ideas, Japan and Germany not only met the targets of the environmental regulations but also succeeded in increasing productivity to higher levels than before. The success led to the competitiveness of companies in Japan and Germany.

In 1978, as the air pollution caused by automobile exhaust gas became serious, the Japanese version of the Muskie Law to reduce NOx emissions by 90% was enforced. At that time, it was also said that it would be impossible to achieve such a decrease because of the extremely high target value, but Japanese manufacturers worked diligently to realize this. Later, as environmental regulations around the world became stricter, Japanese technology had a major advantage.

Innovations often happen when seeking to achieve a high target that initially seems impossible. The important thing is setting a target that seems impossible at first sight, but not reckless.

The authors' research team made technical predictions concerning air conditioners in 1990 based on the thermodynamic theory and also the heat balance. Air conditioners use heat pump technology for absorbing and releasing heat against the temperature difference between indoors and outdoors. The energy efficiency is represented by the numerical value of the coefficient of performance. Before 1990, the coefficient of performance was 3 because air conditioners could heat up or cool down 3 kW with 1 kW of electricity.

The theoretical limit of the coefficient of performance is calculated as the indoor temperature divided by the temperature difference. The temperature here is absolute temperature in which 273 is added to the temperature in Celsius ($^{\circ}$C). Assuming that the room temperature is 28 $^{\circ}$C and the outside air is 35 $^{\circ}$C, the coefficient of performance will be $(273 + 28) / (35-28) = 43$. In other words, 43 times the heat of the consumed electric energy can be pumped out of the room by the heat pump. This number represents is the theoretical value, which is large compared with the value of 3 before 1990. Thus, there is considerable room for innovation here.

In 1990, our team predicted the coefficient of performance in 2050 to be 12. A quadrupling of the coefficient of performance means that the efficiency will be quadrupled. Manufacturers' technicians strongly objected, saying that it was impossible to achieve the coefficient of 12, and it was completely ignored by the former Ministry of International Trade and Industry (currently the Ministry of Economy, Trade, and Industry). However, we had confidence in achieving the coefficient of 12. For the prediction, we even considered improving the efficiency of magnets of motors used in compressors of air conditioners. Since the compressors at that time were consuming twice the amount of electricity as the theoretical value, we thought that we could solve it by using a more efficient magnet. In addition, we examined detailed elemental technologies such as fluid dynamics technology and lubricant technology, and judged that 12 was an appropriate level of possible achievement rather than the theoretical value of 43.

The prediction was largely accurate. As of 1990, when the coefficient of performance was 3, the value of 12 in Vision 2050 seemed to be an extremely high target, but it increased to 7 in 2010, and has steadily evolved thereafter. In response to this trend, the Ministry of Economy, Trade, and Industry also raised the target value for 2050 to 12. Looking at the circumstances, the importance of applying theory and comprehensive and realistic technological forecasts is obvious, rather than merely being a prediction based on the past. In order to induce innovation, high but reasonable target setting is necessary.

4.2.2 Enhance Efficiency with an Energy Management System

In spring of 2016, full liberalization of electricity retailing began in Japan. In order to switch electric power companies, it is necessary to introduce a smart meter equipped with a communication function, and exchange of conventional electric power meters with the smart meter has been progressing. Since the Ministry of Economy, Trade, and Industry has launched a policy to introduce this meter to all households in the 2020s, its uptake will continue to spread in the future.

The smart meter measures the amount of electricity used every 30 min and can transmit data using the communication function. Electric power companies will not only need to conduct meter reading work, but also utilize big data to develop diversified fee structures and adjust supply and demand by pricing.

The advantage for the customer is to be able to view energy consumption trends in real time. Information on smart meters can be linked with the Home Energy Management System (HEMS). The HEMS is a mechanism to optimize energy consumption by connecting all equipment related to electricity in the home such as lighting equipment, air conditioning equipment, photovoltaic power generation systems, household fuel cells, water heaters, and electric cars. By visualizing this information, customers will be able to view whether their electricity consumption is high compared with other households or whether the amount of consumption at this time is larger than expected. As a result, it is expected that the energy conservation awareness will increase in each household. This information is also useful when the customer selects a price plan.

The HEMS also has a great feature in that it can be used to control the connected equipment. It is possible to optimize the whole by balancing the lighting and air conditioning so that a comfortable environment can be maintained or by switching the household electric appliances that consume a large amount of electric power to eco operation according to the remaining amount of power. It is also possible to operate "EcoCute" with solar power to acquire hot water and electricity at the same time, as well as to manage energy consumption and conserve energy.

Similar to HEMS, there are BEMS (Building Energy Management System) for buildings and FEMS (Factory Energy Management System) for factories as systems that optimize energy use in the system. Furthermore, CEMS (Cluster/Community Energy Management System) can also be established in areas where HEMS, BEMS, and FEMS are gathered.

There are several problems to overcome to make these systems popular. Among these, the biggest problem is price.

In a project that was carried out by Toyota City in Aichi Prefecture upon the city's receiving designation from the Ministry of Economy, Trade, and Industry as a Next-Generation Energy and Social System Demonstration Area, remarkable results have emerged with regard to energy self-sufficiency.

As a result of the distribution of smart houses with energy creation (3.6 kW solar power generation), energy saving, equipment for energy storage equipment (including EV) and control by HEMS in the residential area of the city, energy consump-

Table 4.5 Reducing heating energy to 1/12

Coefficient of performance of air conditioners						
Before 1990	1997	2004	2006	2010	Vision 2050	Theoretical value
3	4	5	6	7	12	43

Source: "The Revitalization for Japan"

tion at home and EV 70% of the respondents became energy self-sufficient. In other words, the possibility of self-sufficiency of household and transport energy has come into perspective.

The market price of HEMS is around ¥200,000. Although there are some differences depending on the accompanying function, even the inexpensive systems exceed ¥100,000. Considering that personal computers and tablet devices can be purchased even at around tens of thousands of yen, the current cost of HEMS is too expensive. This is largely because companies try to recover development costs in the early stages of dissemination.

However, there are also successful product examples such as heat-tech products by UNIQLO and Toray's partnership which realized a speedup of product dissemination by pricing based on mass dissemination to the market from the beginning. This was made possible by collaboration between a large-scale production process that realizes low cost and a large-scale sales network that sells at a low price (Table 4.5).

The dissemination of HEMS is one of the keys to creating a low-carbon society. Companies should have the courage to consider the supply system and develop pricing based on the premise that they will be mass disseminated from the beginning.

4.2.3   Japan Should Compete with High-Added-Value Items

Let's think about the costs that are the key to the dissemination of products and services from another angle.

Table 4.6 shows the selling price and the partial cost per weight of the products that are used daily. Most of the weight of products is made up of basic materials such as steel, aluminum, glass, and plastic. The cost of the materials is about ¥0.1 to ¥0.2 per gram. The selling price of mass-produced home appliances such as electric fans, washing machines, and refrigerators is about 5 to 10 times the cost of the materials. The price of light cars and trucks is only five times to ten times as much as the cost of the materials.

The closer the material cost and the selling price, the lower the added value. From this point of view, solar cell modules are becoming commodity products like fans and trucks, and it is clear that price competition is intense.

Li-ion batteries are also considered to be more expensive than other storage batteries, but in terms of weight unit price, there is no major difference from dry batteries. Therefore, Sony and Nissan decided to withdraw from the Li-ion battery

Table 4.6 Selling price of products by weight

Product	Selling price (Yen/g)
Mobile phone	100 ~ 600
Jet aircraft (B787)	100
Watch	50 ~ 3,000 ~
Large gas turbine	15 ~ 30
Personal computer	10 ~ 30
Lithium-ion battery	6
Television	4 ~ 10
Passenger car	1 (light vehicle) ~ 6 (Lexus)
Refrigerator	1
Truck	0.8 ~ 2
Washing machine	0.8
Dry battery	0.7 ~ 4
Electric fan	0.7 ~ 3
Solar cell module	0.7 ~ 2

Source: Created based on materials from the Center for Low-Carbon Society Strategy, Japan Science and Technology Agency

business. It was judged that it is no longer necessary for companies to invest in research and development of their own exclusive products, and all they have to do is to procure general purpose goods from the market. However, the battery market still has plenty of room for technical development. For those who understand this market and technology situation and make appropriate business decisions, it would be possible to expand the industry.

Meanwhile, large-sized gas turbines represent an example of a high value-added product whose selling price per product weight is much larger than the cost. Gas turbines have advanced product design, material development, manufacturing technology, and their price is more than 100 times the price of the materials. The price of an aircraft with a more complicated system than a simple product such as a battery has an additional digit.

In the future low-carbon society, along with the hardware product industry that creates products such as robots, sensors, computers, and medical equipment, progress can be expected in the services industry for freely utilizing these hardware products. We believe that Japan should compete in such high-added-value areas.

In terms of improving the dissemination of HEMS, we should apply a model in which the basic system sets the price assuming mass dissemination and each company competes with high-added-value services that make use of HEMS data. For example, the early mobile phone market had a model for popularizing products by lowering hardware prices and raising revenues from fees. The HEMS data provide the user with the time they get up or sleep, and also actions related to the use of electricity, such as cooking, bathing, and hobbies. Handling of personal information is difficult, but if settled, a high value-added service will be built.

4.2.4 Increased Sophistication of Demand Forecasting by Utilizing Big Data

Approximately 300 million tons of fossil resources are imported to Japan. Roughly, 60% go to oil refineries, 25% to power plants, 5% to gas companies, and the remainder represents coal for iron making. Of these, oil refineries, power plants, and gas companies do not aim to consume energy themselves. It is the daily life (transportation, households, and businesses) and monozukuri sectors mentioned earlier that actually consume energy. Oil refineries, power plants, and gas companies convert energy into different forms so that consumers can use it easily, so they are called the energy conversion sector.

It is preferable to convert fossil resources without loss in the energy conversion sector, but some energy is consumed here. In the case of thermal power generation, the turbine is rotated by the power of steam generated by burning fuel, but since steam does not flow unless one direction is of lower pressure, there is a step in which steam is liquefied in a condenser. About 60% of the total energy from fuel burning is released to the sea and the atmosphere here. Japan's thermal power boasts the world's top efficiency, but the loss is still not small. Loss occurs other than in thermal power generation. Power generation loss, daily life, and monozukuri each account for about one third of the total energy consumption.

Besides improving power generation efficiency, power consumption prediction is also important to reduce power generation losses. Electric power companies adjust the amount of electricity to be shipped by predicting the power consumption based on past power consumption conditions and weather forecasts (temperature predictions). If the amount of electricity generated is larger than the amount consumed, we will discard unused electricity wastefully, but if the electricity generation is less than the amount consumed we will face supply shortage, and society will fall into turmoil. Therefore, electric power companies must constantly generate electricity to a level higher than the predicted power consumption. It is troublesome not to be able to store the electricity once generated.

A research team of Center for Low Carbon Society Strategy (LCS) investigated the power consumption prediction in areas served by TEPCO and the weather forecast data of the meteorological observatory and found that there is an error of several percent. In order to eliminate this error, the accuracy of both the weather forecast model and the power consumption prediction model must be increased. If the weather forecast model improves, it will be easier to set up the power generation plan, and if the accuracy of the demand forecast increases, it will be easier to control power transmission. If prediction accuracy improves, it is expected to reduce energy loss and reduce cost in hundreds of billions of yen, so we should develop highly accurate prediction models using artificial intelligence (AI) and big data (Fig. 4.5).

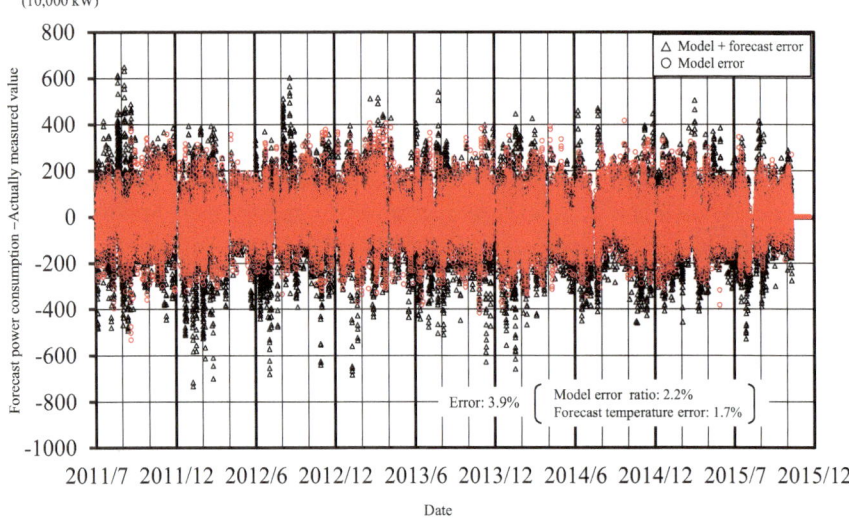

Fig. 4.5 Forecast power consumption error in grid area of Tokyo Electric Power Company. (Source: Created based on materials from the Center for Low-Carbon Society Strategy, Japan Science and Technology Agency)

4.2.5   The Possibility of Carbon Pricing

A "stranded asset" is an asset whose value decreases as the social environment changes. In 2013, the concept of a stranded asset was announced by the Carbon Tracker Initiative, a non-profit think tank, and the London School of Economics, and fossil fuel resources were considered stranded assets.

For the development of coal, petroleum, and natural gas, companies invest $674 billion a year, spending more than $6 trillion over a period of 10 years. In order to keep the rise in average global temperature within 2 °C before the industrial revolution, we cannot utilize 60 to 80% of the recoverable reserves of these fossil fuel resources, and most of the investment in these resources will become useless. In other words, they would be considered stranded assets.

In 2015, Oxford University showed the possibility that many of the domestic coal-fired power plants in Japan will become stranded assets. This is because there are plans to build an excessive number of power plants, considering the competition with other power sources and the renewal cost of existing coal-fired power. The value of the power plants is ¥7 trillion to ¥9 trillion. If a coal-fired power plant becomes a stranded asset, it means that the electric companies made a major mistake in management decision-making.

The problem lies in the cost of CO_2 emissions not being taken into account in companies' decision-making. Decisions on the kind of power station to build should be made after carbon pricing (i.e., after determining the cost of CO_2 that would be discharged by the development).

The setting of an appropriate CO_2 price is difficult to do immediately. However, it is an important viewpoint with regard to realizing a low-carbon society.

Vision 2050 assumes that progress in low carbonization naturally progresses as technology advances, but if there is a mechanism under which low carbonization takes on value, the speed of change can be increased. Hybrid cars are a typical example of products with value in terms of low carbonization. Despite the high price of the car itself, the purchaser judges that it is more profitable in total, if fuel costs are also taken into consideration.

With regard to the carbon tax (taxation on CO_2 emissions), it is better to tax at the stage when resources are shipped, as far as possible. This is because it is easier to set a tax rate according to the amount of CO_2 emissions, such as 3 times of the shipping cost for coal, 2 times for petroleum, or 1.2 times for gas. Alternatively, there is a way to change the tax rate according to the contribution of each resource to GDP. To date, 38 countries have introduced some type of carbon pricing and Japan can learn from many case examples.

Bibliography

Inoue T, Yamada K (2017) Economic evaluation toward zero CO_2 emission power generation system after 2050 in Japan. Energy Procedia 142:2761–2766

Yamada K, Komiyama H (2002) Photovoltaic engineering. Nikkei BP, Tokyo

Chapter 5
Low-Carbon Society in 2050

5.1 Low Carbon Power Supply Systems in 2050

5.1.1 Means to Achieve Low Carbonization

At COP 21, Japan set a long-term goal of reducing CO_2 emissions by 80% of 2013 levels by 2050. As we have seen in Chap. 4, we will try to reduce carbon from a range of angles, but in order to achieve the high target of 80% reduction, we must also optimize the configuration of the power supply used for power generation.

The following methods can be considered to significantly reduce CO_2 emissions from the power supply sector:

1. Decrease the amount of power generation.
2. Increase power generation efficiency.
3. Increase the ratio of power generation by natural gas fuel, which has low CO_2 emissions per calorific value.
4. Add CCS to fossil fuel power generation.
5. Increase the proportion of power generation by renewable energy.
6. Increase the proportion of nuclear power generation.

Let's look at these in order. With regard to the amount of electricity generated (1), measures such as switching to highly efficient electric appliances has progressed after the Great East Japan Earthquake, and it has already declined by about 10%. The annual power generation amount is within 1000 TWh, and this trend will continue in the future. If the annual rate drops by 1%, the consumption of electricity will be about 650 TWh per year in 2050, 30% lower than the current rate. For the direct use of heat and components of fossil fuels (e.g., automobiles and hot water supply), replacement with electricity will continue in order to save energy, so the demand will increase, but we believe that annual amount of electricity generated will not exceed 800 TWh.

© The Author(s) 2018
H. Komiyama, K. Yamada, *New Vision 2050*, Science for Sustainable Societies,
https://doi.org/10.1007/978-4-431-56623-6_5

Table 5.1 Electricity generation cost (Yen/kWh) and CO_2 emissions (g-CO_2/kWh)

	2013		2030		2050	
	Cost	CO_2	Cost	CO_2	Cost	CO_2
Nuclear power	8.8	20	8.8	20	8.8	20
Hydroelectric power	10.8	11	10.8	11	10.8	11
Coal	7.7	943	7.8	881	7.8	881
LNG	10.8	473	11.4	430	11.8	430
Oil	16.7	738	17.9	738	18.9	738
Solar power	16.0	38	9.5	15	5.7	15
Wind power	14.1	25	10.2	25	10.2	25
Geothermal power	12.5	15	12.5	15	8.0	15
Biomass	33.6	5	10.9	5	10.9	5

Source: Created based on materials from the Center for Low Carbon Society Strategy, Japan Science and Technology Agency

With regard to power generation efficiency (2), solar power generation that will be used in large quantities in the future is expected to have an efficiency of 30%, which is 1.5 times the present value. Even with natural gas power generation, under combined power generation in which fuel cells work together from the same heat source, the power generation efficiency may be nearly 70%, or more than 1.2 times the present value. Regarding (3), natural gas has a low CO_2 emission level of about 50% that of coal, and there is room to increase its usage. However, in a society in which reduction of CO_2 emissions by 80% will be achieved, natural gas will not become the main power source. With regard to CCS in (4), there are difficulties in the uptake of the technology such as economic restrictions and restrictions on site conditions, but there is a possibility that it will be carried out in the future, and it is worth considering. In terms of nuclear power (6), the situation is such that it will not become the main power supply under the current circumstances of low public support.

Considering the above, renewable energy (5), which is expressly stated in Vision 2050, is the option that seems most promising in the future. In the next section, we will discuss the establishment of a power supply sector incorporating a large amount of renewable energy.

5.1.2   80% Reduction and Power Generation Costs

Table 5.1 lists power generation costs and CO_2 emissions per kWh that will become widespread in 2050. To generate the values of renewable energy in 2050, we used the cost of the state of technology in 2030 mentioned in Chap. 4, taking into account the speed of its dissemination. With regard to power supply in 2050, solar power generation and wind power generation will be introduced in large quantities. Since the amount of power generated in the case of both solar and wind power changes

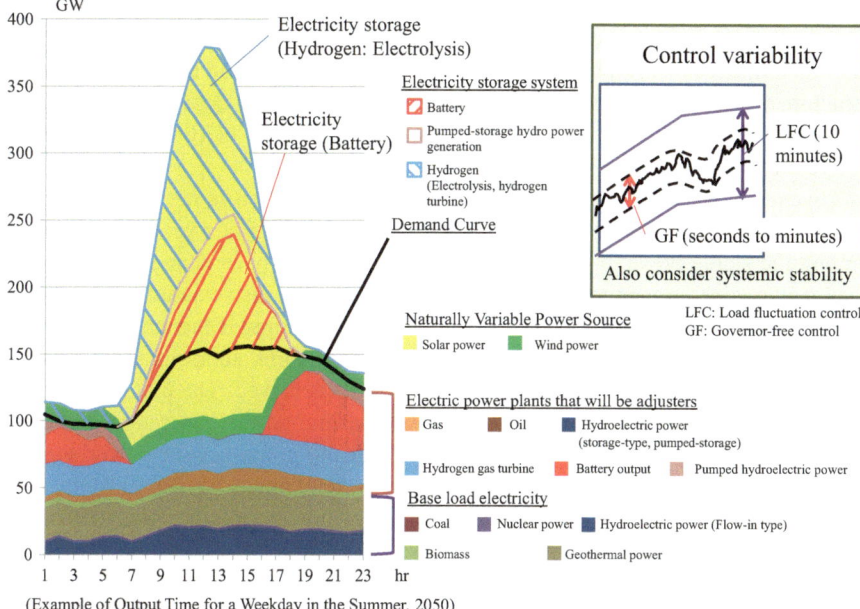

(Example of Output Time for a Weekday in the Summer, 2050)

Fig. 5.1 Summertime power supply and demand storage model. (Source: Created based on materials from the Center for Low Carbon Society Strategy, Japan Science and Technology Agency)

depending on the weather, a power storage system is necessary to achieve a balance between power supply and demand.

Figure 5.1 shows an example of the balance between power supply and demand, focusing on summer weekdays when there are large fluctuations. To achieve balance between power supply and demand in the order of 10 min, introducing a storage battery should suffice. However, taking into account frequency control in the order of several tens of milliseconds as well as response to emergency blackouts, we need a power supply that consists of 50% that uses a rotating machine such as a turbine (New Energy and Industrial Technology Development Organization, "NEDO Renewable Energy Technology White Paper: Issues That Should Be Overcome and Solutions Toward Expanding Dissemination of Renewable Energy, 2nd Edition," 2014, pp. 635 Morikita Publishing Co., Ltd.). Although there is a possibility that it may become 50% or less due to power grid design and development of a new control system, here I have decided to focus on power supplies in which a rotary power generation system accounts for 50% or more.

Stable power sources capable of rotating power generation are mainly thermal power, hydropower, nuclear power, geothermal power, and biomass power generation. Among these, geothermal power generation is a method that generally utilizes the heat of a geothermal reservoir located at a relatively shallow depth underground, but attention is focused on utilization of high temperature rocks deep underground due to concerns about effects on hot spring areas. Hot dry rock (HDR) geothermal power generation (also known as the enhanced geothermal system) is also effective in reducing electricity costs, and development is also progressing in the U.S. It is not so complex system, but more investigation and research on deep underground structure are necessary to develop the technology for practical application. In Japan, it would be possible to complete a power generation system of 100 TWh (13 GW) per year by 2050.

Given that the status of use of nuclear power generation is not foreseeable in 2050, there is considerable uncertainty in the role of nuclear and geothermal power among the stable power sources by 2050. Therefore, we have changed the annual total power generation amount to 600–1200 TWh, and we also examined how much nuclear and geothermal power generation would be required if the amount of electricity generated was reduced. We would also like to think about minimization of costs when a reduction of 80% of CO_2 emissions is achieved. The unit power generation costs for nuclear power and HDR geothermal power generation and CO_2 emissions are almost equal. As a result, however the proportion of power supply from the two power sources changes, the effect on total power cost can still be seen by using the value of the sum of the amount of power generated by these two sources.

5.1.3 Consideration of the Best Power Supply Configuration

Table 5.2 shows the comparison of power generation costs calculated by changing the annual power generation amount, power supply configuration, and CO_2 reduction rate. As mentioned in the previous section, since power consumption is estimated to be 650 TWh per year by 2050, the annual demand for electricity has been set with 700 TWh as the reference value in Table 5.2, considering that this is a highly feasible numerical value. In addition, we calculated the cost of power generation by setting various patterns from 600 TWh, which is slightly less than the reference value, to 1200 TWh, which is higher than the reference value.

Cases 1, 2, and 4 are examples in which the annual demand for electricity has been set between 600 and 1000 TWh, and the amount of CO_2 emissions has been reduced by 80%. In these cases, nuclear power generation and HDR geothermal power generation are not included, but if the demand is less than 800 TWh, the cost would be about ¥11/kWh in all cases. If the demand falls slightly below the current level, it is possible to reduce the electricity cost below the current level of ¥12/kWh even without nuclear power and HDR thermal power generation.

In contrast, if the demand reaches 1000 TWh as in Case 5, the cost rises sharply to ¥18/kWh. Even power savings of about 20% will be effective in reducing cost and the amount of CO_2 emissions under such a scenario.

In Case 3, the demand for power is the same as in Case 2, but the target for reducing CO_2 emissions has been increased to 90%. The power cost at this point is high at ¥16/kWh.

Table 5.2 Cost by power supply configuration and percentage of CO_2 reduction in 2050

Case		1	2	3	4	5	6	7
Demand for electricity (TWh)		600	700	700	800	1,000	1,000	1,200
Annually generated electricity (TWh)	Nuclear power	0	0	0	0	0	0	0
	Hydroelectric power	130	130	130	130	130	130	130
	Coal	100	28	0	0	0	14	0
	LNG	19	166	97	224	211	190	330
	Solar power	373	398	544	494	746	500	746
	Wind power	8	8	7	9	38	24	124
	Geothermal power	12	12	12	12	12	12	12
	Geothermal power (hot dry rock)	0	0	0	0	0	200	0
	Biomass	30	31	31	31	31	22	31
	Total	672	772	822	900	1,169	1,092	1,373
Battery capacity (GWh)		423	466	666	631	780	621	795
Amount of hydrogen use (TWh)		0	0	53	0	118	0	119
Cost of generating electricity (Yen/kWh)		10.8	10.8	16.1	11.0	18.4	9.9	17.3
Annual CO_2 (Mt-CO_2)		113	113	57	113	113	113	170
Rate of CO_2 reduction (Cf. 2013)		80%	80%	90%	80%	80%	80%	70%

Source: Created based on materials from the Center for Low Carbon Society Strategy, Japan Science and Technology Agency

In Case 6, the demand for power and CO_2 emission reduction target are the same as in Case 5, but the HDR geothermal power generation of 200 TWh is included. By increasing stable renewable energy, cheap coal-fired power can be used and storage battery capacity is decreased compared to Case 5, so the power generation cost is the cheapest.

Also, as shown in Case 7, if the demand for power is as high as 1200 TWh, the power generation cost will substantially increase, even if the CO_2 reduction rate is reduced to 70%.

It should be noted that the value of total power generation is higher than the demand for power because it is calculated so that the cost is minimal. In order to set surplus power generation to 0, storage batteries and hydrogen power generation are used together, but they are high cost, so it is cheaper to discard the surplus. Even if solar power and wind power generation used to produce hydrogen cost ¥0, electrolysis equipment and transportation incur expenses. In Cases 3, 5, and 7, the power generation costs are high, but hydrogen power generation is included as stable power supply in all cases.

From the above, it has been clarified that if the power supply configuration is as presented in Cases 1, 2, and 4, the target of reducing CO_2 emissions by 80% in 2050 can be achieved without increasing the power generation cost and without introducing nuclear power and HDR geothermal power generation.

5.2   Reducing Carbon in Major Fields

5.2.1   Value-Added Industry and Low Carbon

What approach should we take to create a bright low carbon society (a platinum society) using low-carbon power that achieved a reduction in CO_2 emissions by 80%? Let us consider using the data for 2013, which is the base year for CO_2 emissions in Japan in COP 21.

Since CO_2 emissions in 2050 will be reduced by 80%, this represents a mass of 250 million tons (deducting 1003 million tons from 1254 million tons). If the annual economic growth rate is assumed to be 0.4%, the nominal GDP value is estimated to be ¥600 trillion, which is about ¥80 trillion more than it is now. In terms of CO_2 emissions per ¥1 trillion of GDP, the amount of CO_2 emissions must be drastically reduced from the current 2.4 million tons to 420,000 tons.

Table 5.3 illustrates the amount of CO_2 emissions, added value, and CO_2 emissions per ¥1 trillion of value-added for each industry sector in 2013. The power sector emits a large amount of CO_2, accounting for 45% of the total emissions (567 million tons), but since most of the emissions are allocated to each industry sector, they are not represented in this figure. However, as shown in the previous section, it is possible to reduce emissions from the power sector by 80% (454 million tons).

In the demand sector, households emit 224 million tons of CO_2, which is equivalent to 18% of the total. By saving energy in the household and with a new power supply configuration, we estimate that CO_2 can be reduced by 179 million tons (of which power usage accounts for 129 million tons), which is equivalent to 80% of CO_2 emissions in this sector. The transportation sector can also reduce 196 million tons of CO_2 (of which power usage accounts for 8 million tons) which is also 80% of CO_2 emissions. In total, a reduction of 375 million tons can be achieved.

Therefore, the amount of CO_2 emissions to be reduced in other industrial sectors, excluding emissions due to power usage (where reduction by 317 million tons is possible), CO_2 emissions could be reduced by a total of 311 million tons.

5.2.2   The Ideal State of the Steel Industry

The steel sector emits the largest amount of CO_2 in the industrial sector. As mentioned in Chap. 2, 40–70 billion tons of new steel is needed globally in the future. Some of the required new steel could possibly be manufactured in Japan even in 2050, considering the high quality of Japanese steel products and high-level of energy saving technology. Production of steel for export will increase CO_2 emissions, but it will be difficult for steel consuming countries to bear this burden. Thus, when we consider reductions in CO_2 emissions by 80% in 2050, it is necessary to consider the increase in CO_2 emissions due to exports.

Table 5.3 Volume of CO_2 emissions by industrial sector (2013)

Industries and households	Total CO_2 emissions (Millions of tons)	Amount of value-added (Trillions of yen)	Volume of CO_2 emissions per one billion worth of value-added
Agriculture, forestry, and fishery industries	4.2	6.45	654
Mining industry, etc.	2.5	0.35	7,164
Construction	12.3	31.48	390
Food and drinks manufacturing industry	21.1	15.39	1,369
Textile industry	12.2	2.26	5,396
Wood products, furniture, and other industries	2.6	0.70	3,658
Pulp, paper, and paper products manufacturing	23.7	2.37	10,013
Printing and related industries	2.8	2.68	1,048
Chemical industry	82.0	8.00	10,248
Petroleum products and coal products manufacturing industries	3.0	4.37	683
Plastic, rubber, and leather products manufacturing industries	10.0	1.46	6,850
Cement, sheet glass, and limestone manufacturing industries	45.7	3.10	14,765
Iron and steel industry	199.8	5.95	33,578
Non-ferrous metal manufacturing industry	8.0	1.60	5,004
Metal products manufacturing	7.3	5.46	1,333
General machinery and devices manufacturing industries	2.3	11.23	205
Production machinery and devices manufacturing industries	4.0	2.85	1,411
Industrial machinery and devices manufacturing industries	4.0	1.51	2,675
Electric circuits for electric parts and devices manufacturing industries	10.4	5.60	1,865
Telecommunications machinery and devices manufacturing industries	2.3	1.28	1,818
Transportation machinery and devices manufacturing industries	14.8	13.00	1,140
Machine manufacturing and other products	4.0	1.93	2,082

(continued)

Table 5.3 (continued)

Industries and households	Total CO$_2$ emissions (Millions of tons)	Amount of value-added (Trillions of yen)	Volume of CO$_2$ emissions per one billion worth of value-added
Other manufacturing industries	1.5	5.26	292
Electricity, gas, heat, and water supply industries	10.6	9.12	1,167
Telecommunications industry	20.6	28.83	716
Transportation and postal industries (excluding privately owned vehicles)	259.2	26.55	9,760
Wholesale and retail industries	63.3	70.97	893
Finance and insurance industries	2.5	23.48	107
Real estate and rental industries	18.5	66.35	279
Academic research, specialized, and technology services	5.6	0.28	20,263
Hospitality and food services	48.8	14.08	3,470
Life-related services and entertainment industries	33.1	13.59	2,439
Education and learning assistance industry	17.8	0.68	26,121
Medicine and welfare	29.1	36.79	790
Multiple-services industries	0.6	27.06	23
Other service industries	35.0	11.20	3,126
Official duties	4.5	59.83	76
Households	224.1	0	–
Total	1,254	523	2,400

Source: Created based on materials from the Center for Low Carbon Society Strategy, Japan Science and Technology Agency

Since the amount of CO$_2$ emissions from the steel sector in 2013 was 200 million tons per year, the amount of emissions will be 40 million tons for the same level of production if we achieve an 80% reduction in 2050. Among the current CO$_2$ emissions, the amount of emissions attributed to power generation is 51 million tons, which can be reduced by 80% if we use future low carbon power supplies. The issue is how to reduce other emissions from within the steel production process.

Table 5.4 is an illustration of the steel production in 2050 when such a significant reduction is achieved. There are two major methods of steel production: the Blast Furnace Method using iron ore as a raw material and the Electric Furnace Method using iron scrap as a raw material. Table 5.4 displays simulated images of the steel industry in 2013 and 2050. The supply and demand of steel by each manufacturing method, and the amount of CO$_2$ emitted in correlation with steel supply are presented in the figure. The situation in 2050 is based on the following five assumptions.

Table 5.4 Present and future demand for iron and steel, examples of CO_2 emissions

		2013 (190 million ton (Mt) of CO_2 emissions)	2050 (40 million ton of CO_2 at 80% reduction)	
			Case 1	Case 2
		Iron and steel (CO_2 emissions, Mt)	Iron and steel (CO_2 emissions, Mt)	Iron and steel (CO_2 emissions, Mt)
Supply		121 (200)	80 (64)	67 (40)
Details	Blast furnace method	86 (185)	30 (54)	17 (30)
	Electric furnace method	25 (15)	50 (10)	50 (10)
	Import	10	0	0
Demand		121	80	67
Details	Domestic	58	40	37
	Exported steels, etc.	42	20	10
	Exported products	21	20	20

Source: Created based on materials from the Center for Low Carbon Society Strategy, Japan Science and Technology Agency

1. Use of future electric power that reduces CO_2 emissions by 80%.
2. Because of the increase in recycled steel, the production volume of the Electric Furnace Method will be double the current volume.
3. The Electric Furnace Method will achieve energy saving by 20%.
4. The production rate by the Blast Furnace Method will be significantly reduced.
5. Steel export volume will be significantly reduced.

Based on the above, we calculated two case examples. It should be noted that the steel produced by the current Electric Furnace Method cannot be used for certain products such as thin automobile plates and large-diameter pipes. Currently, development is underway to produce iron and steel materials with properties suitable for such purposes.

In Case 1, total supply is assumed to be 80 million tons, which is a reduction of 30% from the current level, on the premise that domestic demand will decrease and export of steel materials will be narrowed down to high value-added products in the future. In addition, steel production by the Electric Furnace Method is assumed to be doubled to 50 million tons in 2050. The Electric Furnace Method has an energy conservation rate of 20%. In this case, the total CO_2 emissions are 64 million tons, which fall short of the target of 80% reduction from 2013 levels.

In Case 2, export volume of steel from the demand side is further lowered. The steel production by the Blast Furnace Method was 17 million tons, making the total production 67 million tons. This will result in CO_2 emissions of 40 million tons, achieving an 80% reduction from 2013 levels.

In order to achieve an 80% reduction under the conditions of Case 1, the CO_2 emissions must be further reduced by 24 million tons. To this end, it is conceivable to combine the Blast Furnace Method with CCS. Assuming that CCS cost will be ¥4000 per ton of CO_2, the annual cost will be about ¥96 billion, and the cost increase per ton of steel will be about ¥3200, if the cost increase is allocated to the whole Blast Furnace Method. It may be possible to manufacture high value-added steel corresponding to this cost increase.

Therefore, a more realistic scenario is Case 2, where the Electric Furnace Method using recycled steel is doubled and steel production using the Blast Furnace Method is reduced by approximately 80%. If we can change the structure of steel production in this way, reduction of CO_2 emissions by 80% is possible. In this case, the level of unit CO_2 emissions using the Blast Furnace Method is assumed to be the same as the current level. If improvements to the manufacturing process continue, it will be possible to increase the production quantity of steel using the Blast Furnace Method commensurate with that progress.

5.3 Reducing CO_2 Emissions by 80% Across Japan

5.3.1 Low Carbonization by Sector in 2050

Reducing CO_2 emissions by 80% in the three sectors (households, transportation, and steel) is not easy; however, it is feasible. These emissions account for 53% of Japan's total emissions, and thus CO_2 emissions from other sectors must also be reduced by 80%.

Table 5.5 illustrates the CO_2 emissions by sector in the base year 2013 and in 2050, by which CO_2 emissions are to be reduced by 80%. Out of the total emissions in other sectors (586 million tons of CO_2/year), 60% is derived from electricity, so this can be reduced by 80% in the future as previously discussed. The focus is the remaining 40%, involving reduction of emissions due to the use of fossil fuels from 235 million tons to 47 million tons. This can be realized with energy saving and an increase in electricity usage ratio. While expanding economic activities, it is

Table 5.5 CO_2 emissions of the 3 sectors (households, iron and steel, and transportation) for 2013 and for 2050 when 80% reduction will have been realized

| | | CO_2 emissions (Mt) | | |
		Electricity	Other than electricity	Total
The 3 sectors	2013	216	452	668
	2050	43	90	133
Other sectors	2013	351	235	586
	2050	70	47	117
Total	2013	567	687	1,254
	2050	113	137	250

Source: Created based on materials from the Center for Low Carbon Society Strategy, Japan Science and Technology Agency

necessary to promote low-carbon technologies or systems and to accelerate changes in the industrial structure through innovation.

Tables 5.6a, 5.6b, and 5.6c presents an overall perspective of the situation. Let's consider Japan's CO$_2$ emissions and value-added by industry sector in 2013 based on each of these plots.

Table 5.6a Amount of value-added and percentage by sector (2013)

Order	Industry, etc.	Amount of value-added (Trillions of yen)	Percentage (%)	Volume of CO$_2$ emissions (kt)	Percentage (%)	Volume of CO$_2$ emissions per one billion yen worth of value-added (t)
1	Wholesale and retail industries	70.97	13.6%	63,346	5.1%	893
2	Real estate and rental industry	66.35	12.7%	18,513	1.5%	279
3	Official duties	59.83	11.4%	4,546	0.4%	76
4	Medical and welfare	36.79	7.0%	29,060	2.3%	790
5	Construction	31.48	6.0%	12,264	1.0%	390
6	Telecommunications industry	28.83	5.5%	20,629	1.6%	716
7	Multiple-services business	27.06	5.2%	626	0.0%	23
8	Transportation and postal industries + Transportation fuels	26.55	5.1%	259,183	20.7%	9,760
9	Finance and insurance industries	23.48	4.5%	2,507	0.2%	107
10	Food and drinks manufacturing industries	15.39	2.9%	21,076	1.7%	1,369
11	Hospitality and food service industries	14.08	2.7%	48,847	3.9%	3,470
12	Life-related services and entertainment industries	13.59	2.6%	33,145	2.6%	2,439
13	Transportation machinery and devices manufacturing industries	13.00	2.5%	14,823	1.2%	1,140
14	General machinery and devices manufacturing industries	11.23	2.1%	2,299	0.2%	205
15	Other service industries	11.20	2.1%	35,030	2.8%	3,126

(continued)

Table 5.6a (continued)

Order	Industry, etc.	Amount of value-added (Trillions of yen)	Percentage (%)	Volume of CO_2 emissions (kt)	Percentage (%)	Volume of CO_2 emissions per one billion yen worth of value-added (t)
16	Electricity, gas, heat, and water supply industries	9.12	1.7%	10,640	0.8%	1,167
17	Chemical industry	8.00	1.5%	82,018	6.5%	10,248
18	Agriculture, forestry, and fishery industries	6.45	1.2%	4,219	0.3%	654
19	Iron and steel industry	5.95	1.1%	199,814	15.9%	33,578
20	Electric circuits for electric parts and devices manufacturing industries	5.60	1.1%	10,448	0.8%	1,865
21	Metal products manufacturing industries	5.46	1.0%	7,280	0.6%	1,333
22	Other manufacturing industries	5.26	1.0%	1,535	0.1%	292
23	Petroleum products and coal products manufacturing industries	4.37	0.8%	2,982	0.2%	683
24	Cement, sheet glass, and limestone manufacturing industries	3.10	0.6%	45,741	3.6%	14,765
25	Production machinery and devices manufacturing industries	2.85	0.5%	4,026	0.3%	1,411
26	Printing and related industries	2.68	0.5%	2,805	0.2%	1,048
27	Pulp, paper, and paper products manufacturing industry	2.37	0.5%	23,718	1.9%	10,013
28	Textile industry	2.26	0.4%	12,185	1.0%	5,396
29	Machine manufacturing and other products	1.93	0.4%	4,014	0.3%	2,082
30	Non-ferrous metal manufacturing industry	1.60	0.3%	8,016	0.6%	5,004

(continued)

Table 5.6a (continued)

Order	Industry, etc.	Amount of value-added (Trillions of yen)	Percentage (%)	Volume of CO$_2$ emissions (kt)	Percentage (%)	Volume of CO$_2$ emissions per one billion yen worth of value-added (t)
31	Industrial machinery and devices manufacturing industry	1.51	0.3%	4,035	0.3%	2,675
32	Plastic, rubber, and leather products manufacturing industries	1.46	0.3%	10,000	0.8%	6,850
33	Telecommunications machinery and devices manufacturing industry	1.28	0.2%	2,319	0.2%	1,818
34	Wood products, furniture, and other industries	0.70	0.1%	2,563	0.2%	3,658
35	Education and learning assistance industry	0.68	0.1%	17,772	1.4%	26,121
36	Mining industry, etc.	0.35	0.1%	2,489	0.2%	7,164
37	Academic research, specialized, and technology services	0.28	0.1%	5,591	0.4%	20,263
38	Households (all energies)			224,098	17.9%	
	Total	523.08		1,254,201	100.0%	2,398

Table 5.6a shows value-added by industry sector and their shares. Total value-added of the top five industries accounts for 51% of the GDP, but the total CO$_2$ emissions from these industries is merely 10% of the total. In contrast, Table 5.6b, where industries are sorted in descending order of CO$_2$ emissions, the sum of emissions from the top five industries account for 52% of the total, but the sum of value-added is merely 24% of the GDP. Furthermore, according to Table 5.6c, where industries are sorted by CO$_2$ emissions per ¥1 trillion of value-added, the sum of industries between 26th and 37th from the lowest value-added accounts for 71% of the GDP. However, the sum of CO$_2$ emissions remains at only 13% of the total.

Just because an industry's value-added is large, it does not mean that the amount of CO$_2$ emitted is necessarily large, or vice versa. Needless to say, the most desirable combination is large value-added with small CO$_2$ emissions. Building on this, the major issues to be addressed are as follows:

Table 5.6b   Volume of CO_2 emissions by sector (2013)

Order	Industry, etc.	Volume of CO_2 emissions (kt)	Percentage (%)	Amount of value-added (Trillions of yen)	Percentage (%)	Volume of CO_2 emissions per 1 billion yen worth of value-added (Tons)
1	Transportation and postal industries + Transportation fuels	259,183	20.7%	26.55	5.1%	9,760
2	Iron and steel industry	199,814	15.9%	5.95	1.1%	33,578
3	Chemical industry	82,018	6.5%	8.00	1.5%	10,248
4	Wholesale and retail industries	63,346	5.1%	70.97	13.6%	893
5	Hospitality and food service industries	48,847	3.9%	14.08	2.7%	3,470
6	Cement, sheet glass, and limestone manufacturing industries	45,741	3.6%	3.10	0.6%	14,765
7	Other service industries	35,030	2.8%	11.20	2.1%	3,126
8	Life-related services and entertainment industry	33,145	2.6%	13.59	2.6%	2,439
9	Medical and welfare	29,060	2.3%	36.79	7.0%	790
10	Pulp, paper, and paper products manufacturing industries	23,718	1.9%	2.37	0.5%	10,013
11	Food and drinks manufacturing industry	21,076	1.7%	15.39	2.9%	1,369
12	Telecommunications industry	20,629	1.6%	28.83	5.5%	716
13	Real estate and rental industries	18,513	1.5%	66.35	12.7%	279
14	Education and learning assistance industry	17,772	1.4%	0.68	0.1%	26,121
15	Transportation machinery and devices manufacturing industry	14,823	1.2%	13.00	2.5%	1,140
16	Construction	12,264	1.0%	31.48	6.0%	390
17	Textile industry	12,185	1.0%	2.26	0.4%	5,396
18	Electricity, gas, heat, and water supply industries	10,640	0.8%	9.12	1.7%	1,167

(continued)

Table 5.6b (continued)

Order	Industry, etc.	Volume of CO$_2$ emissions (kt)	Percentage (%)	Amount of value-added (Trillions of yen)	Percentage (%)	Volume of CO$_2$ emissions per 1 billion yen worth of value-added (Tons)
19	Electric circuits for electric parts and devices manufacturing projects	10,448	0.8%	5.60	1.1%	1,865
20	Plastic, rubber, and leather products manufacturing	10,000	0.8%	1.46	0.3%	6,850
21	Non-ferrous metal manufacturing	8,016	0.6%	1.60	0.3%	5,004
22	Metal products manufacturing industry	7,280	0.6%	5.46	1.0%	1,333
23	Academic research, specialized, and technology services industry	5,591	0.4%	0.28	0.1%	20,263
24	Official duties	4,546	0.4%	59.83	11.4%	76
25	Agriculture, forestry, and fishery industries	4,219	0.3%	6.45	1.2%	654
26	Industrial machinery and devices manufacturing industry	4,035	0.3%	1.51	0.3%	2,675
27	Production machinery and devices manufacturing industry	4,026	0.3%	2.85	0.5%	1,411
28	Machine manufacturing and other products	4,014	0.3%	1.93	0.4%	2,082
29	Petroleum products and coal products manufacturing	2,982	0.2%	4.37	0.8%	683
30	Printing and related industries	2,805	0.2%	2.68	0.5%	1,048
31	Wood products, furniture, and other industries	2,563	0.2%	0.70	0.1%	3,658
32	Finance and insurance industries	2,507	0.2%	23.48	4.5%	107
33	Mining industry, etc.	2,489	0.2%	0.35	0.1%	7,164

(continued)

Table 5.6b (continued)

Order	Industry, etc.	Volume of CO_2 emissions (kt)	Percentage (%)	Amount of value-added (Trillions of yen)	Percentage (%)	Volume of CO_2 emissions per 1 billion yen worth of value-added (Tons)
34	Telecommunications machinery and devices manufacturing industries	2,319	0.2%	1.28	0.2%	1,818
35	General machinery and devices manufacturing industry	2,299	0.2%	11.23	2.1%	205
36	Other manufacturing industries	1,535	0.1%	5.26	1.0%	292
37	Multiple-services business	626	0.0%	27.06	5.2%	23
38	Households (all energies)	224,098	17.9%			
	Total	1,254,201	100.0%	523.08		2,398

1. Structural change of industries to services with low CO_2 emissions per value-added and further reduction of CO_2 emissions.
2. As exemplified in the high-emission steel industry, the intra-industry system change is required centered on recycled products.
3. Promotion of energy conservation in all industrial fields.
4. Introduction of low carbon power supply systems through establishment of a large amount of renewable energy production.
5. Creation of new low-carbon industries.
6. Development of new functional materials supporting the above changes.

5.3.2 Value-Added by Industry and CO_2 Emissions in 2050

As we have seen so far, reducing 80% of CO_2 emission by 2050 in the power sector as well as the other three sectors (household, steel, and transport), which together account for 81% of total CO_2 emissions, is now within reach. If we can reduce the remaining 19% of the total CO_2 emissions by 80%, we can achieve an overall reduction of 80% in 2050. Since we can see this much improvement at this stage, we have also applied the same numbers to related industries. The remaining amount is small, so we decided on an emission reduction of roughly 50 to 90% (on average 80%) considering the situation of each industry.

Quantitatively forecasting the value-added and industrial structure of each industrial sector in 2050 is more difficult than projecting the CO_2 emissions. In this esti-

Table 5.6c Sector-specific volume of CO_2 emission per value-added (2013)

Order	Industry, etc.	Volume of CO_2 emissions per 1 billion yen worth of value-added (Tons)	Amount of value-added (Trillions of yen)	Percentage (%)	Volume of CO_2 emissions (kt)	Percentage (%)
1	Iron and steel industry	33,578	5.95	1.1%	199,814	15.9%
2	Education and learning assistance industry	26,121	0.68	0.1%	17,772	1.4%
3	Academic research, specialization, and technology services	20,263	0.28	0.1%	5,591	0.4%
4	Cement, sheet glass, and limestone manufacturing industries	14,765	3.10	0.6%	45,741	3.6%
5	Chemical industry	10,248	8.00	1.5%	82,018	6.5%
6	Pulp, paper, and paper products manufacturing industries	10,013	2.37	0.5%	23,718	1.9%
7	Transportation and postal industries + transportation fuels	9,760	26.55	5.1%	259,183	20.7%
8	Mining industry, etc.	7,164	0.35	0.1%	2,489	0.2%
9	Plastic, rubber, and leather products manufacturing industries	6,850	1.46	0.3%	10,000	0.8%
10	Textile industry	5,396	2.26	0.4%	12,185	1.0%
11	Non-ferrous metal manufacturing industries	5,004	1.60	0.3%	8,016	0.6%
12	Wood products, furniture, and other industries	3,658	0.70	0.1%	2,563	0.2%
13	Hospitality and food service industries	3,470	14.08	2.7%	48,847	3.9%
14	Other service industries	3,126	11.20	2.1%	35,030	2.8%
15	Industrial machinery and devices manufacturing	2,675	1.51	0.3%	4,035	0.3%
16	Life-related services and entertainment industry	2,439	13.59	2.6%	33,145	2.6%

(continued)

Table 5.6c (continued)

Order	Industry, etc.	Volume of CO_2 emissions per 1 billion yen worth of value-added (Tons)	Amount of value-added (Trillions of yen)	Percentage (%)	Volume of CO_2 emissions (kt)	Percentage (%)
17	Machine manufacturing and other products	2,082	1.93	0.4%	4,014	0.3%
18	Electric circuits for electric parts and devices manufacturing industry	1,865	5.60	1.1%	10,448	0.8%
19	Telecommunications machinery and devices manufacturing industry	1,818	1.28	0.2%	2,319	0.2%
20	Production machinery and devices manufacturing industry	1,411	2.85	0.5%	4,026	0.3%
21	Food and drinks manufacturing industry	1,369	15.39	2.9%	21,076	1.7%
22	Metal products manufacturing industry	1,333	5.46	1.0%	7,280	0.6%
23	Electricity, gas, heat, and water supply industries	1,167	9.12	1.7%	10,640	0.8%
24	Transportation machinery and devices manufacturing industry	1,140	13.00	2.5%	14,823	1.2%
25	Printing and related industries	1,048	2.68	0.5%	2,805	0.2%
26	Wholesale and retail industries	893	70.97	13.6%	63,346	5.1%
27	Medical and welfare	790	36.79	7.0%	29,060	2.3%
28	Telecommunications industry	716	28.83	5.5%	20,629	1.6%
29	Petroleum products and coal products manufacturing	683	4.37	0.8%	2,982	0.2%
30	Agriculture, forestry, and fishery industries	654	6.45	1.2%	4,219	0.3%
31	Construction industry	390	31.48	6.0%	12,264	1.0%
32	Other manufacturing industries	292	5.26	1.0%	1,535	0.1%
33	Real estate and rental industry	279	66.35	12.7%	18,513	1.5%

(continued)

Table 5.6c (continued)

Order	Industry, etc.	Volume of CO$_2$ emissions per 1 billion yen worth of value-added (Tons)	Amount of value-added (Trillions of yen)	Percentage (%)	Volume of CO$_2$ emissions (kt)	Percentage (%)
34	General machinery and devices manufacturing industry	205	11.23	2.1%	2,299	0.2%
35	Finance and insurance industries	107	23.48	4.5%	2,507	0.2%
36	Official duties	76	59.83	11.4%	4,546	0.4%
37	Multiple-services business	23	27.06	5.2%	626	0.0%
38	Households (all energies)				224,098	17.9%
	Total	2,398	523.08		1,254,201	100.0%

Source: Created based on materials from the Center for Low Carbon Society Strategy, Japan Science and Technology Agency

mation, we assumed the GDP as value-added in 2050 to be ¥600 trillion, and then distributed it after considering the situation of each industry.

The estimation is based on the following premises. On the supply side, labor productivity is improved due to advances in technologies such as AI, sensors, robots, and computers, as well as sophistication of systems that make use of them. In addition, development of low-carbon, high-performance products and new processes through the development of new materials will proceed, and production of new materials, parts, and structural materials will also expand. However, new needs will be created on the demand side. Although there may be changes within service industries such as information and communications, medicine and welfare, accommodation and catering, education, wholesale and retail, and finance and insurance, their size will expand as a whole.

Overall, technological innovation for low carbonization will progress, and a change in industrial structure accompanying expansion of economic activity will occur. Table 5.7 shows the current state (2013) and the future image (2050) of industries that are expected to grow and expand in the future. Value-added will increase from ¥190 trillion to ¥262 trillion in seven industries, mainly in the services industry, such as information and telecommunications, medical care and welfare, and accommodation and catering, while CO$_2$ emissions will decrease from 200 million tons to 55 million tons. These industries cover most of their energy with electricity, and it is possible to further decrease CO$_2$ emissions by development of energy saving of electric drive equipment. Therefore, it can be said that it is easier to promote low carbonization in these industries compared with the manufacturing industry.

Table 5.7 Examples of industries that can expand by 2050

		Value-added (trillion Yen)		CO_2 emissions (G t)		CO_2 Emissions (t) per billion Yen of value-added	
		2013	2050	2013	2050	2013	2050
1	Information and Communication	29	50	21	10	720	200
2	Medicine and Welfare	37	52	29	8	790	150
3	Hospitality and Food Services	14	25	49	12	3,500	480
4	Wholesale and Retail	71	80	63	7	890	90
5	Financial and Insurance	24	30	2.5	0.3	110	9
6	Life Related, Services, and Entertainment	14	20	33	10	2,400	500
7	Education and Research Technology Services, etc.	1	5	23	8	23,000	1,600
Total		190	262	221	55	(Average: 1,160)	(Average: 210)

Source: Created based on materials from the Center for Low Carbon Society Strategy, Japan Science and Technology Agency

In Japan, which will have realized a reduction of 80% of CO_2 emissions by 2050, the total CO_2 emissions would be 250 million tons and the GDP ¥600 trillion. The CO_2 emissions per ¥1 billion value-added (hereinafter referred to as "CO_2 emissions per value-added") is expected to be 420 tons, which is a significant reduction from the current 2400 tons.

Table 5.7 shows the average emission of industries per value-added, which is calculated to decrease from current 1160 to 210 tons.

Let's look at each of these industries individually. Although the value-added of "information and communication" is estimated to be 1.7 times larger than current level, this simulation is based on a conservative assumption where the annual average growth rate is 1.5%, so there is a possibility of further expansion.

Similarly, under the "medical care and welfare" category, given the recent progress in medical equipment, medical technology, treatment modalities, and the amount of wealth possessed by the elderly people, the number of people willing to pay extra for receiving treatment is expected to increase. If we have a comfortable hospital environment in combination with high medical technology in Japan, the demand from overseas will also increase, which could double the current value-added. If Japan's state-of-the-art medical system is constantly adopted and flexible management becomes possible, this value will be further increased.

In the medical field, the percentage of electricity use is large, so even the low carbonization of electricity alone will lead to a considerably large low carbon industry, and it is still essential to achieve high insulation in hospital buildings and further low carbonization of various medical equipment. The amount of power required by the heavy particle radiotherapy equipment used for cancer treatment is as high as 3000 kW with one unit. The electricity cost accounts for a considerable part of the total treatment cost (¥3 million per person), and the amount of CO_2 emissions is large, yet in principle, it is possible to save a substantial amount of electricity. For low carbonization of not only heavy particle radiotherapy equipment but also for the medical and welfare sector to be achieved, it is important to develop a system using various high functional materials. This is also a suitable field to leverage Japan's strengths.

By incorporating the themes discussed above into concrete issues and working to solve each issue, it is possible to achieve both targets of economic expansion and low carbonization. Given the accelerated changes in future society, we consider that enhancing the "education and research technology services" sector is a particularly important approach. To be able to respond to the ever-changing society, it is necessary to have an educational system that allows everyone to acquire new knowledge and ways of thinking and continue learning. Such a system will inevitably grow as an industry. Furthermore, the expansion of research service sectors such as society and engineering or science to drive this social change and to discover the seeds such as new technologies and systems for a new society will similarly advance. There is a possibility that the value-added in these sectors may grow beyond the scope of these estimations.

5.3.3   Image of CO_2 Emissions and Changes in GDP in all Industries

Lastly, let us consider low carbonization of all industries. The CO_2 emissions data per value-added in 2013 and 2050 are interspersed, and Fig. 1.13 is graphically presented once again (Fig. 5.2). The vertical axis shows CO_2 emissions per unit value-added, while the horizontal axis shows value-added.

In Table 5.6a, 5.6b and 5.6c, the CO_2 emissions of households that do not produce value-added are ranked number 38 of domestic sector. Fig. 5.2 is provided to demonstrate the relationship between value-added of the industrial sector and CO_2 emissions, by allocating the total CO_2 emissions of the household sector into 62.8 million tons to the petroleum and coal product manufacturing industry, and into 161.3 million tons to the electricity, gas, and water supply industry.

This is how Fig. 5.2 should be interpreted. In the case of 2013, the value-added (GDP) on the horizontal axis is the highest at ¥523 trillion, and the lowest plot on the vertical axis where CO_2 emissions per unit value-added is 23,000 tons/¥1 trillion is the services industry. In Table 5.6c, it is ranked at 37th, the lowest plot. If ¥27 trillion of value-added of the services industry is deducted, the remaining GDP is ¥496 trillion. The plot at ¥496 trillion on the horizontal axis and at 76,000 tons/¥1 trillion on the vertical axis in Fig. 5.2 is the public sector.

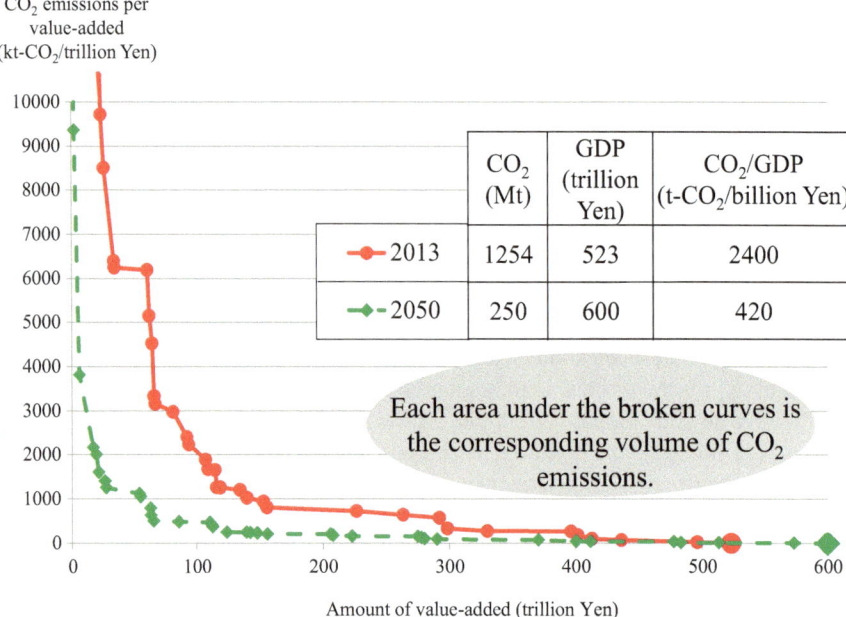

Fig. 5.2 CO_2 emissions and value-added by industry. (Source: Created based on materials from the Center for Low Carbon Society Strategy, Japan Science and Technology Agency)

Since the area inside the line connecting the plots indicates the total CO_2 emissions, low carbonization means minimizing this area. Therefore, to create a bright low carbon society, it is important to increase the value-added while lowering the CO_2 emissions per value-added.

Chapter 6
Platinum Industry and a New Society

6.1 What Is a Platinum Society?

6.1.1 Per Capita GDP and Average Life Expectancy

The problems we are facing now, which are characteristic of developed countries, have been caused by the rapid advances of the twentieth century. As shown in the diagram (Fig. 1: Trajectory of Sudden Expansion in Human Development in the twentieth Century), average life expectancy worldwide and per capita GDP portray the exact same arc. From the changes shown, we can get a solid sense of how turbulent the twentieth century was.

Until the Industrial Revolution, the rise in both average life expectancy and per capita GDP was gradual. However, both show a steep rise from the beginning of the twentieth century and within the space of 100 years, per capita GDP spiked six times higher. Average life expectancy quickly reached beyond 70 years of age and in countries famous for longevity such as Japan, average life expectancy entered into the 80 years of age bracket. The reason why average life expectancy rose is that because of the development of agriculture and industry, there was an increase in the number of people who could easily obtain food. Far from that, overeating and obesity became social problems in Japan and diseases such as dementia have increased along with the rise in longevity.

Examining closely the issues characteristic of developed countries, it all seems to come down to global warming and the aging of society. The material culture that expanded rapidly in the twentieth century triggered an increase in energy consumption as well as CO_2 emissions, causing global warming, and as a result, climate change is now a reality.

Social structure also underwent major changes. As mentioned earlier, average life expectancy worldwide is more than 70 years. Developed countries are now dealing with problems such as old-age pensions, medical expenses, and cases of dementia. Japan has covered costs for old-age pensions and medical expenses at social

© The Author(s) 2018
H. Komiyama, K. Yamada, *New Vision 2050*, Science for Sustainable Societies,
https://doi.org/10.1007/978-4-431-56623-6_6

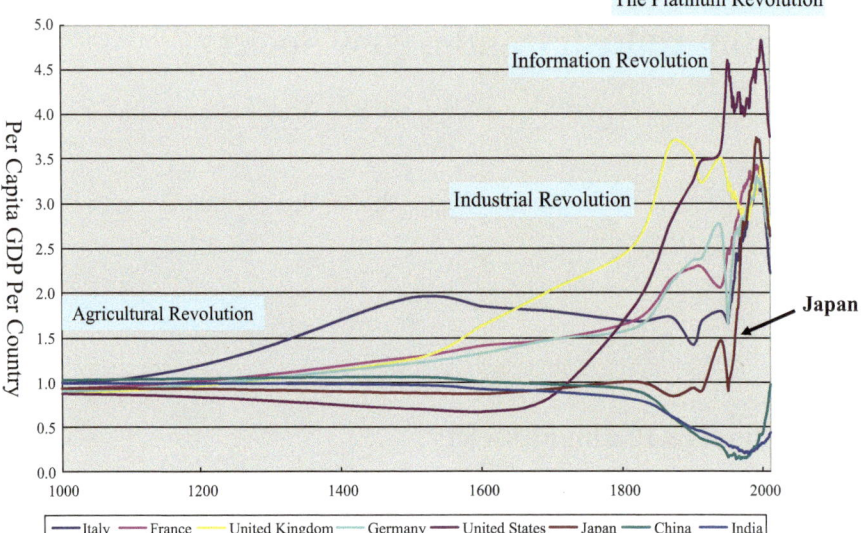

Fig. 6.1 Change in per capita GDP of major countries with per capita GDP in the world as the Standard. (Created by the Author. Source: Angus Maddison: Statistics on World Population, GDP and Per Capita GDP, 1-2008 Conference Board Total Economy Database™, January 2012, http://www.conference-board.org/data/economydatabse/)

security costs. However, it is self-evident that as the population ages, these costs will become excessive. Furthermore, through the saturation of man-made objects, demand for something new will decline and demand will be mainly for replacement products, causing economic growth to slow down even further. Things will not keep going properly by simply adhering to a conventional line of thinking.

The aging of the population can be seen as the realization of longevity. Life expectancy at the beginning of the twentieth century was merely 31 years. Humankind has been successful in doubling life expectancy over a 100 years span, making the dream of longevity come true. Although some elderly people require nursing care, others are still quite healthy. Many people can still be active in society and be independent with just a little support. The experience and knowledge that older people have acquired over a long number of years can be put to use in the field of education. Preventative medicine for avoiding the need for care will probably become a new industry. The aging of society is not just comprised of risks but also contains some kind of opportunities as well.

These issues are common to developed countries including Japan, and developing nations are bound to experience them at some point. Out of all the nations faced with challenges, if Japan can take the initiative in facing up to these problems and possibly take the first step towards resolution, then Japan will become a leading country in resolving societal problems.

Figure 6.1 shows the changes in per capita GDP in major powers, with the global per capita GDP as the standard. Humankind has negotiated various major turning points thus far, such as the Agricultural Revolution, the Industrial Revolution, and the Information Revolution. Each time economic growth resulted. Undoubtedly the opportunity for Japan to create a new social model has arrived.

6.1.2   From Quantitative Sufficiency to Qualitative Sufficiency

In the twentieth century, humankind continually kept going in their race toward material affluence, health and longevity. An impasse in material culture and the aging population are contemporary problems but such problems have arisen because humankind made long cherished dreams come true.

From now on, as the next kind of affluence, quality should be sought over quantity. From 2010, I have been proposing the Platinum Society in which citizens who are satisfied in quantity in turn seek a high-quality society. Platinum contains various kinds of luminance including green for ecology, silver for health, and bright red for IT. I coined it to mean a lifestyle of a level higher than others. In the twentieth century, quantitative sufficiency was sought after and advances in science and technology made that possible. The twentieth century was a golden century for humankind. If that was a gleaming golden age then the twenty-first century, when qualitative sufficiency is being achieved, must be a glorious platinum century that shines brightly. In other words, the Platinum Society is a model of growth in a mature society, and as mentioned in the introduction, has the following views of the world.

1. No insecurity about resources, energies, etc.
2. No pollution, but maintenance of sustainability in the earth's environment
3. Living in harmony with a diverse and beautiful nature
4. Health and maintaining self-reliance long term
5. Opportunities to participate lifelong in society
6. Lifelong growth
7. Employment opportunities
8. Rich in both cultural and qualitative terms

These are items to be taken up not only by developed countries, including Japan, but by countries all over the world. This is a pioneering model that hold true for all countries. On the negative side, developed countries generated issues such as pollution and lifestyle-related diseases in the process of seeking quantitative sufficiency. Developing countries need not take the same path as taken by developed countries. Instead, they should aspire to a Platinum Society (Fig. 6.2).

The Platinum Society does not conflict with a low-carbon society or the Vision 2050 mentioned earlier.

The Platinum Society Network, of which I am the founder and chairman, presents awards every year to entities from all over Japan that are implementing excellent initiatives. The inaugural Grand Prize in 2013 was presented to the town of Ama, on an island in Shimane Prefecture, where a revitalization project involving the local high school led to the revitalization of the island. The inaugural Award for Excellence was presented to Kamikatsu town in Tokushima prefecture. The town is famous for its *Irodori* (meaning bright colors) Project.

Fig. 6.2 Amachō, known for high school revival, received First Place in the first Grand Prize for the Platinum Competition

Kamikatsu town originally produced mandarin oranges. However, in 1981 the mandarin orchards suffered devastating damage from abnormally cold weather. In an effort to recover financially, Kamikatsu residents started up a project of selling to restaurants the flowers and leaves they found growing naturally in hilly areas of the town. Although they encountered difficulties in the beginning, the project became a resounding success after the town's residents persevered by analyzing customers' needs and conducting market surveys, developing new products and skillfully incorporating IT.

Nowadays, old ladies living in Kamikatsu use computers to sell their goods to restaurants in cities. The elderly residents of Kamikatsu are all enjoying good health with work giving meaning to their everyday lives. There are hardly any elderly who are bedridden in Kamikatsu.

With a community base like this, Kamikatsu has begun its "Sanitation System Inspired by a Zero Waste Policy." Although the town achieved success with its *Irodori* Project, the local municipality does not have much leeway in its finances with a population of just under 1700. The residents have come together to implement a strict system of separating rubbish into 34 different categories for recycling. This enabled the town to greatly reduce its rubbish processing costs. The town has teamed up with residents and Lixil Corporation to promote the demonstration of a new wastewater purification system, thus playing a part in preserving the environment.

Many hints for resolving issues faced by developed countries are to be found in the series of initiatives implemented by Kamikatsu town.

6.1.3 An Island (Ama-cho) that Increased the Number of Children Attending School Despite a Declining Birthrate

The Dozen group of islands, located in Oki district in Shimane prefecture, is another region that has overcome issues.

A declining birthrate and aging population are both proceeding at a fast pace in the Dozen island area. In 2007 the number of children aged 15 was 51, less than half of what it had been at the peak period. The Shimane Prefectural Oki-Dozen Senior High School, located in the town of Ama, in the center of the Dozen area, had only 28 new students enter in 2008, reaching a new low that threatened the school's existence.

The residents of Ama helped to launch the Oki-Dozen Senior High School Appeal Project. All the island's residents sought to make the school recognized as being very attractive in the field of education, and they enacted a strategy to stop the younger generation from leaving the island and to attract families with young children, and former residents back to the island.

The high school set up a community development course and prepared a unique curriculum that includes community life studies, work experiences at local businesses, and studies to resolve issues through collaboration with the local community. Using interaction with high school students as a starting point, the local community is aiming to revitalize the region by making use of local resources. The sightseeing plan *Hito-tsunagi* (connecting people), which was put together mainly by Dozen High School students, did an amazing job of receiving first prize in a national sightseeing plan contest for high school students.

As well as the creation of substantial content education-wise, a dormitory was set up to enable students who are not from the island to attend the school. Activities to attract students for *Shima-ryugaku* (Studying on the Island) are being promoted. The catchy Japanese name has been receiving attention from the media and in 2012, enough applicants were received to boost the number of classes, something unprecedented. Nowadays nearly half of the new students entering the school are not local island residents.

This project at Ama is unique in the way that it connected the school and the local community. Measures aimed at stopping the decline in population are usually focused on developing industry and creating employment. However, as the town of Ama is poor in industrial resources, the community decided to do their best by using education as their own special feature. The town hopes to nurture the kind of people who can find out what the issues are on their own initiative and think of a way to resolve those issues while networking with other various people. This can viewed as a model for creating human resources in a problem-solving developed country.

Actually, from the perspective of regional development, this project has even greater significance. If a community loses its senior high school, the teenagers will leave their hometown when they graduate from junior high school. Upon graduation from senior high school, the teenagers will probably leave their hometown too, but if they spend those extra 3 years on the island when they are at such an impressionable age, there is sure to be a big difference in their feelings and a deepening of their affection for their hometown. In actual fact, quite a few students who graduate from Dozen Senior High School leave the island saying that they will return after they have finished their university education. Right now many places around Japan are trying to copy the successful Dozen Senior High School initiative. This is because the initiative has proved its worth.

6.1.4 Contributing to Lowering Carbon in Asia from Actual Experiences (Kitakyushu)

Kitakyushu in Fukuoka prefecture has long been a city with heavy and chemical industries, focused mainly on steelworks. In the 1960s the city experienced serious pollution issues. The city's residents demanded to have their blue sky restored and various countermeasures were implemented. As a result, by the beginning of the 1990s the city was able to overcome its pollution issues. This result was brought about by collaboration amongst various stakeholders including the residents of Kitakyushu, the city office, corporations and the national government. The city of Kitakyushu became the focus of attention from other cities in Japan and also cities overseas which were experiencing similar problems. At the World Summit on Sustainable Development held in Johannesburg in 2002, the "Kitakyushu Initiative" was incorporated in the Plan of Implementation, for municipalities around the world to learn from the experiences of Kitakyushu. In 2013, the Organization for Economic Co-operation and Development (OECD) selected Kitakyushu City as part of its Green City Program along with Paris, Stockholm, and Chicago (Fig. 6.3).

In 2010, Kitakyushu city established the "Kitakyushu Asian Center for Low Carbon Society." As well as aiming to vitalize the regional economy through eco-business, the Center acts as a core facility for contributing to green growth which makes economic growth compatible with environmental conservation in Asia. The "Kitakyushu Model" is composed of five categories that have a major impact on urban environments. The categories are management of waste, town water, sewage, and energy, as well as environmental conservation and traffic control. These categories are then further systemized and structuralized with three main elements: roles; processes; and outcomes. This is all managed in a cloud-based database. As it was anticipated that this would be used in countries outside of Japan, from the beginning the database was set up in English and then translated into Japanese and Chinese so that it can be used in three different languages. The database has the records of various actions that have been taken from the past through to the present. It also includes mistakes from the past which serve as lessons for the future.

The Asian Center for Low Carbon Society supports cities in various Asian countries including Vietnam, Indonesia, Cambodia and Thailand, in resolving urban environmental problems. As a model environmental city, Kitakyushu may be more famous in Asia rather than in Japan. In the future, the Center intends to cover not only low carbon but also sustainable economy and ageing.

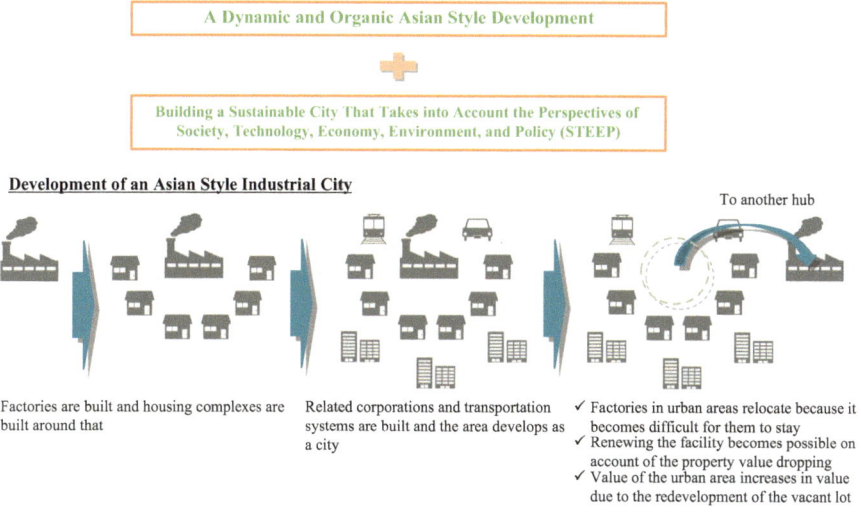

Fig. 6.3 Special characteristics of the Kita-Kyushu Model: Asian dynamism. The Kita-Kyushu model has at its foundation a global standard platform of urban renewal, but it is also a "model of an environmentalized city that is compatible with an Asian industrial city" which has included the developmental trajectory of Kita-Kyushu City which is both an industrial city and a future-facing environmental city. This makes it possible to take an Asian Style approach that cannot be accommodated by a Western Style one which presumes a top-down zoning policy. (Source: MI Consulting)

6.1.5 Leadership that Achieved a Miracle (Yanedan)

Resources and problems that have to be resolved are not all the same for various regions. The important thing is to have a vision and work towards achieving that vision.

In the city of Kanoya, in Kagoshima Prefecture, is the mountainous area of Yanagidani, commonly known as Yanedan. This area was also experiencing the aging of its population as well as a declining birthrate. However, the community was able to revitalize the area over a period of ten-something years and their success is referred to as the Yanedan Miracle. The central figure in this is Tetsuro Toyoshige who was appointed head of the local community center in 1996. Mr. Toyoshige felt a sense of crisis when he considered that in the population of 300 people, the aging ratio stood at 40% and also hardly any community activities were being undertaken. He unfurled his vision of not depending on government but for Yanedan residents to generate an independent source of funds and to revitalize the local community. Then action was taken (Fig. 6.4).

By using abandoned farmland provided by local residents, Mr. Toyoshige started cultivating sweet potatoes with the help of local youth. He then set about establishing the Yanedan brand of *shochu*, or Japanese traditional spirit distilled from sweet potatoes. Red chili peppers are also grown and exported to Korea. In this way, Yanedan was able to secure an independent source of income. The residents received permission to use land owned by the village and they opened the Wakuwaku Sports

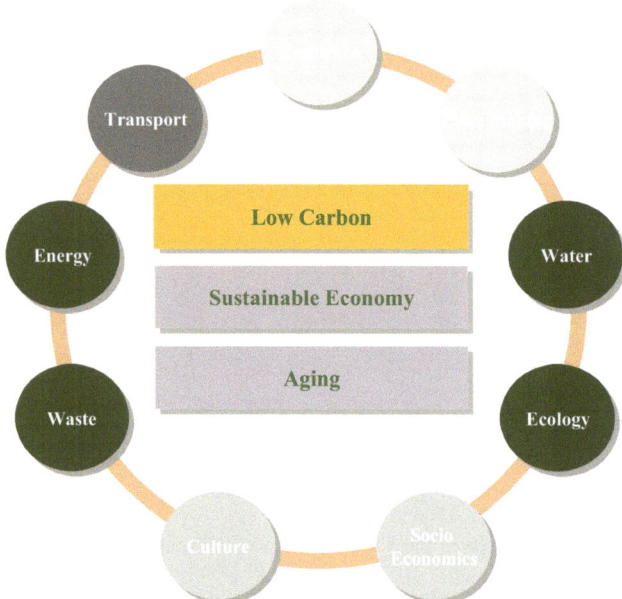

Fig. 6.4 Expandability of the model: The theme of sustainable possibility. (Source: MI Consulting) We are currently organizing the 4 areas of waste management, water and sewer management, energy management, and environmental conservation
We are scheduled to expand through incorporating elements described below in order to further our efforts to build a comprehensive and sustainable city

Park complete with handmade sports equipment. This contributed to promoting the health of the village's residents and the medical expenses for residents of 75 years or older is 40% less than other communities in Kanoya city.

Although the revitalization was triggered by one person who took on the role of leader, nowadays each resident actively takes part in leadership and participating in community activities. In 2007, Mr. Toyoshige established the *Furusato Zoyo Juku*, a learning center for fostering municipal workers who can become leaders involved in revitalizing the community. Three years ago, the *Furusato Zoyo Super Juku* was established for training *Furusato Zoyo Juku* graduates into elite leaders to revitalize the community, thus further broadening community activities.

6.1.6 *Realizing a Vision in a Megalopolis (Futakotamagawa)*

Stories about regional revitalization often occur in isolated areas but urban areas have their own problems too. Futakotamagawa, located beside the Tokyu Electric Railway Line, is home to the Tamagawa Takashimaya Shopping Center, the first suburban shopping center opened by a Japanese department store. Development centered around the shopping center and it was a popular area. It is the destiny of cities to continue to make progress in order to keep on being appealing. The

Fig. 6.5 Image of the growth spiral of a creative city. (Source: Tokyu Corporation)

Futakotamagawa redevelopment project began in 1982 and this mammoth project was finally completed last year after a total of 33 years.

In town planning, not is only development of hardware needed but also software is essential too. To that end, the Creative City Consortium (CCC) was established in 2010 (Fig. 6.5). The executive committee members include Tokyu Corporation, Culture Convenience Club Co. Ltd., and Dai Nippon Printing Co. Ltd., which supports regional revitalization activities throughout Japan.

With Futakotamagawa as a model area, CCC is composed of members from different types of businesses, as well as creative artists and academics, to create an advanced example of a new urban environment that goes beyond the limitations of business categories, where various kinds of activities can be developed in order to realize social systems and workstyles, and lifestyles. The strengths lie in how the mechanism can fuse together the knowledge and expertise of the participating corporations and so on, making it an appropriate venue for the experimental adoption of new ideas. A win-win relationship was forged with stakeholders including local corporations and residents. Here just taking the pick of good points from certain corporations was not permitted and corporations firmly entrenched in community activities were welcomed with open arms. Participating corporations are required to prepare themselves to interact with local stakeholders over the long term.

Five years since its establishment, CCC has expanded its activities to cover the so-called Platinum Triangle, an area connecting Futakotamagawa, Shibuya, and Jiyugaoka. In the Platinum Triangle reside a large number of the "creative class" of people, those who can generate new value, and new innovations will be created with a major consumer market being formed which will be on the receiving end of those innovations. This means that ideas can lead to business opportunities and people will have the chance to realize their dreams in this area.

Fig. 6.6 The public running tour of segway® Personal transporter. (Photos from Tokyu Corporation)

One example of outcomes by CCC is the Segway Tour in Futakotamagawa, the first of its kind wherein participants can ride Segways on public roads in Tokyo (Fig. 6.6). This is the Quomo Project, started up by CCC with mobility as its theme. With the cooperation of Setagaya Ward and the Futakotamagawa Traffic and Environment Cleanup Promotion Council, tours began in April 2016 by utilizing measures for exceptional cases under the Ministry of Economy, Trade and Industry's System of Special Arrangements for Corporate Field Tests. Keeping in mind the Olympic and Paralympic Games Tokyo 2020, CCC is conscious of urban development activities aimed at the next generation and creates and disseminates actual examples by using methods including new technologies and the easing of regulations.

In the future, CCC intends to expand its place for activities from Futakotamagawa to the Platinum Triangle and create innovations necessary for new sustainable urban development in the megalopolis of Tokyo. Tokyo is the driving force for Japanese innovations and the Platinum Triangle is expected to be the engine of growth for this.

We are now facing many problems. However, it is true that through the hard efforts of our predecessors, we have acquired various goods, information, and means of transportation enabling the realization of a society with long life expectancy. Self-sufficiency in resources, low carbon, overcoming pollution and coexisting in harmony with nature, good health and self-reliance, lifelong growth, various options, freedom of participation…Let's work towards the Platinum Society that will realize all of this.

6.2 Towards Becoming a Nation Self-Sufficient in Resources

6.2.1 Making a Self-Sufficiency Rate of 70% a Reality with Vision 2050

The Platinum Society is a society wherein people can be proud and shine in life as well as it being a society that is rich in quality.

One part of the concrete picture of the vision is a nation that is self-sufficient in resources. Japan is said to be a resource-poor country. Japan depends on imports for most of its energy and earns foreign currency after importing raw materials, including minerals, processing them and then exporting those processed products. However, that model has its limitations.

So far just 10% of the world's population in developed countries has monopolized industry and the remaining 90% in the other countries had no choice but to sell primary resources. The developed countries bought the primary resources cheaply and by selling to the world at expensive prices, they were able to achieve economic growth. However, this structure of relationships is already starting to fall apart. The countries that depended on exporting primary resources have started up their own industries. Products are flooding the world and competition over prices is happening, narrowing the gap between resources and products.

As the world is being flooded by yet another industrial revolution, Japan needs to aim towards becoming a nation that is self-sufficient in resources. Being 100% self-sufficient is ideal but in reality, about 70% would be enough.

As discussed in Chap. 1, energy consumption will decrease from now onwards. As proposed in Vision 2050, energy utilization efficiency will improve and if natural energies are expanded, it is possible to reach a rate for energy self-sufficiency of 70% by 2050.

Saturation of man-made objects is also becoming obvious. Technology for recycling is improving and if material recycling systems are better maintained, many materials such as iron and cement, rare metals and rare earths, can be acquired by recycling from urban mines. Aiming for a self-sufficiency rate of 70% for mineral resources is a possibility.

Currently for food, the rate is 40% (on a calorie base) and this too could be increased to 70%.

In regard to water and timber, achieving a rate of 100% is totally within reach. Japan has many sources of water and two thirds of the nation is covered in forests making the potential for timber resources high. Currently, the self-sufficiency rate for timber resources is low at around 25%. However, if forestry is revived, as an industry it would generate somewhere in the vicinity of 5 trillion yen.

Looking at the results above, Japan's self-sufficiency rate for materials would be greatly increased. In line with Vision 2050, achieving a self-sufficiency rate of 70% for materials is indeed possible by 2050.

6.2.2 A Scenario for Reviving Forestry

Despite Japan having an abundance of forest resources, 75%t of the demand for timber relies on imports. About 10 million hectares were afforested for restoration after World War II. However, during the period of economic growth young people left the countryside for the cities and there was a lack of workers in the forestry industry. On top of that, cheaply imported timber became available and domestic timber was buffeted by price competition. Forestry workers aged without any drastic reforms of the industry such as streamlining of business operations or expansion of scale. Due to a labor shortage, forests did not receive adequate care and a vicious cycle occurred with forests becoming wilder as national land conservation worsened. This cycle must be broken.

European timber exporting countries such as Sweden and Austria share common features such as large-scale management, rational mechanization, and creation of supply chains. The logging and carrying costs per cubic meter for these countries are 2000 and 4000 yen respectively, but it is about 8000 yen in Japan. The density of Japan's forestry road network is extremely low, one-tenth and one-fifth of Sweden and Austria, which is one of the reason for high cost of timber in Japan.

Hopes are being pinned on Smart Forestry. It is anticipated that Smart Forestry can overcome various issues with technology and business models. The Platinum Society Network has established a forestry revival working group and is examining exactly what needs to be done by Smart Forestry.

Actually, several advanced cases have already occurred in Japan. For example, in Gunma prefecture forestry associations have cooperated with each other to establish the Shibukawa Sanzai Center where an entire lot is purchased in one lump sum. This center has become the core of the supply chain making it possible to accept various qualities of timber. Another example is the town of Shimokawa in Hokkaido where the whole community tackles the revitalization of forestry (Fig. 6.7). Shimokawa was the first town in Hokkaido to be recognized by the Forest Stewardship Council, which is responsible for certification globally of sustainable forestry control and management and as well as adding higher value to timber, sawdust from the sawmill in the town is used for biomass. It is used in public facilities and this venture is touted as reducing both CO_2 and running costs at the same time.

If supply chains are formed and the use of biomass spreads, the revenue from this can be plowed back into Smart Forestry. If forestry is revitalized, forests will be better looked after and a healthy natural environment can be maintained. This in turn will lead to soil conservation while preventing the occurrence of landslide disasters.

Use of biomass, reduction of CO_2, economic revitalization, strengthening national land, and preservation of the environment and water resources—the benefits from creating supply chains and from smartening the forestry industry are major. That is why the government and corporate groups with massive capital should participate actively and using their size as leverage, should guide Japan's forestry to a strong revival.

Fig. 6.7 Efforts to revitalize forestry & timber industry in town of Shimokawa

6.3   Coexisting in Harmony with Beautiful Nature

6.3.1   A World that Is Comfortable for All Living Things

As mentioned earlier, Japan should aim at becoming a leading country in resolving societal problems. I believe that Japan has the strength to do so. In actual fact, Japan has overcome many difficulties in the past.

When industrialization was proceeding at a fast pace during Japan's period of high economic growth, the natural environment was being polluted at various places throughout the country. Pollution-related diseases such as Minamata disease (from mercury poisoning), Itai-itai disease (from cadmium poisoning), and Yokkaichi asthma (from sulfur-oxide pollution) occurred. Whether it is the United States or Russia or wherever, any areas close to industrial zones are liable to be polluted. However, as Japan is limited in terms of space, houses are quite often located close to factories and people have been directly affected. In light of this situation, the industrial world developed technologies to eliminate toxic substances and thus overcame pollution issues.

If a burden is placed on the natural environment, not only people but all living things will be affected. If forests die and the natural environment falls into ruin, then living things in that area lose their habitat. Many wild animals are in danger of extinction and already some species endemic to Japan have disappeared.

People who are upset at seeing nature in such a state have been trying to make things as they once were long ago and activities for environmental conservation have started up in various areas. These efforts have been rewarded and once again Japanese crested ibises grace the skies of Sado Island in Niigata prefecture, while the white stork has been successfully returned to the wild in Toyooka city in Hyogo Prefecture, and fireflies are once again to be found in Mishima city's Genbei River in Shizuoka prefecture. Each of these habitats is being successfully rehabilitated. In addition, people are connecting these activities to making their areas more appealing. The city of Sado has established the Toki Forest Park which provides services like an environmental study program to learn about the ecology of crested ibises. In Toyooka, a special brand of rice on the market touts that the environment it has been grown in is so beautiful as to be a habitat for white storks. The city of Mishima holds events like its firefly festival which revitalize its tourism.

The prosperity of humankind is only possible through living in harmony with nature. There can be no future for humankind if development that creates a burden for the environment is pursued. A society that lives in harmony with nature is one of the vital elements that make up the Platinum Society.

6.3.2 Initiatives by Corporations for Living in Harmony with Nature

During the years of Japan's economic bubble, corporate support of the arts was popular. However, in the 1990s, attention was focused on environmental problems, and more and more companies conducted environmental activities such as planting trees and conducting cleanups. It is wonderful that there are still companies continuing or even further extending such activities. However, there seems to be a trend for these kinds of activities to be discontinued, especially when such activities are so different to the companies' main type of business. In the same way that corporate support of the arts petered out as the economy slumped, it is only natural that extra activities are cut short when earnings structures worsen.

Environmental conservation activities by corporations should be positive activities that link up with the company's activities, not simply volunteering to help nature that has been damaged. Because such activities will in turn contribute to the company's main business, the company can throw in the necessary resources on a continual basis.

The beverage company Ito En, Ltd., conducts a Program for Revitalizing Tea-Growing Regions. The total area under contract cultivation is 668 hectares while 366 hectares are new tea plantations, making a total of 1034 hectares (as of the end of 2015). Contract cultivation means that farmers grow the tea and Ito En buys the entire amount. New tea plantations refers to large scale tea plantation business operations that use abandoned farmland, and provide various kinds of support to the farmers including technology and use of IT.

For Ito En, revitalizing tea-growing areas is not simply part of their CSR activities but it is advantageous for the company as it is directly linked to the main business, with stable procurement of raw materials for their beverages. That is why the company can continue such activities long term. In addition, management is stabilized through buying the entire contracted amount from farmers and providing technological support. Farming on agricultural land that was once abandoned resolves this issue for the local community and creating jobs adds social value. Thus, this project makes all three parties happy. Ito En recycles used tea leaves into fertilizer and feed. The company was featured in a special article titled "50 Companies That Are Changing The World" in the American business magazine *Fortune* and was ranked at number 18 out of the fifty, the highest for a Japanese company.

I'd like to introduce one more corporate initiative for farming. It is Vegetalia, Inc., a venture business started up in 2010 that promotes smart agriculture in accordance with scientific backing.

Agricultural practices these days came into being the mid-twentieth century and are based on the Green Revolution which has a three-piece set of high-yield varieties of seeds, fertilizers and pesticides. Through this, production increased dramatically but in turn led to issues such as environmental destruction and an increase in abandoned farmland. The biggest issue is that farmers are still sticking to methodology that is more than 50 years old and no innovations have been made. In recent years, with advances made in science, the mechanisms of how plants grow, and how disease and insect damage are caused have become clearer. This means that agricultural production of the bio-harmonized type, which is kinder to the environment, is not impossible.

PaddyWatch, a program developed by Vegetalia, is a tool leading to agricultural innovation through a fusion of science and technology (Fig. 6.8). With PaddyWatch, a sensor is inserted into a rice field to measure the water level, water temperature, temperature and humidity. This data is sent to smartphones or tablets via an app making management by remote control possible. Management of water is very important in growing rice and it is quite a burden to farmers as the amount of time spent on water management is said to make up 25% of rice-farming work time. If this can be monitored, the working efficiency will be boosted.

Moreover, the most effective time to use pesticides and fertilizers can be determined from the data generated so even if the farmer doesn't have much experience, they can carry out the work practices appropriately. Satoshi Koike, president and CEO of Vegetalia, said "Currently, agricultural production relies largely on the experience, intuition and the master skill of experienced farmers but experience can be replaced by scientific backing from plant science, intuition can be replaced by the IoT sensors, while master-skills can be replaced by scientific data and AI."

Methods developed by Vegetalia have been adopted by Niigata city's agricultural reform project which has been designated as a National Strategic Special Zone by the Japanese Government. Niigata is a major farming area in Japan but even so people there have a strong sense of impending crisis for the future of farming. The lack of successors in farming villages coupled with the aging population is of serious concern. As farmers age and their physical strength weakens, the area that they

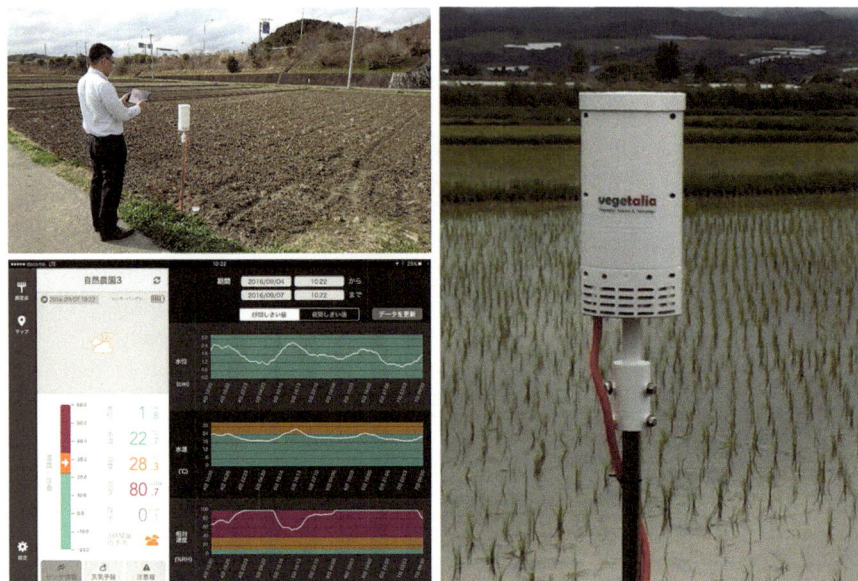

Fig. 6.8 A rice field sensor, "PaddyWatch," uses sensing technology. (Courtesy: Vegetalia)
A sensor attached under the small cylindrical case (shown on the right) measures the water level
and temperature. The data is controlled by a special app and can be accessed on a tablet device or
smartphone (shown on the top left and bottom left).

can cultivate decreases and the amount of unused farm land increases. Once fertile
and productive farm land gets abandoned and the real function of the land is lost,
and the natural environment is laid to waste. Attempts to revive rice fields that have
been dried out and farmland that has been badly neglected is much more trouble
than trying to continue farming and is not a problem that can be solved in a short
time. It bears a striking resemblance to the decline in forestry.

However, agriculture is now facing a new phase through the participation of
corporations. Both for Ito En and Vegetalia, the corporate activities of their own
companies are of primary importance but they could not exist if not for a healthy
natural environment. Their corporate activities are synonymous with environmental
conservation and coexistence with nature. Both are excellent examples of how business and activities contributing to society can be linked together.

Farming must be a commercially practicable industry in order to promote the
participation of corporations. To that end, improving agricultural production by
using ICT is a prerequisite condition. Japanese agriculture is not particularly efficient when compared on a global level. This is especially difficult in respect to
growing rice. The gross output for one hectare of farm land for vegetables is
4,300,000 yen while for rice it is a mere 950,000 yen, making it about one quarter
that for vegetables. This needs to be improved.

Improving productivity will contribute to the promotion of agriculture and a
reduction in abandoned farm land, as well as leading to environmental conservation

of the area. That little sensor standing in a rice field could be a key device in protecting the global environment.

6.4   Good Health and Self-Reliance for a Fulfilling Life

6.4.1   *The Wisdom of Seniors Is a Social Resource*

We should aim for a society where people can enjoy a better Quality of Life (QOL) while maintaining quantitative richness as well. To that end, remaining active for one's entire life is vital.

Most companies have a retirement age of 60 or 65 years of age but the sixties age bracket nowadays is different from the old days. Most people in their sixties are active seniors who enjoy good health while their brains still function well and they are still keen to continue working.

Mayekawa Manufacturing Co., Ltd., a company well-known for its industrial freezing technology, has an unorthodox employment system with its zero retirement policy implemented since 1976. If the employee so desires and if there is a need in the workplace, employment contracts can be renewed annually past the age of sixty enabling the employee to continue working. Work is still full time and social security is still applicable but wages are reduced by about 40% and the workplace also changes.

These employees, who have opted to stay on past 60, are expected to pass on knowledge and techniques they have acquired to younger employees. Younger employees are kept busy everyday with their work duties but when an older employee with experience and leeway joins them, some kind of chemical reaction is said to occur and products that are top class globally can be created through cooperative work between the younger and older generations. Furthermore, when a younger employee hits some kind of wall, they can often make a breakthrough from the intuition that only a veteran employee possesses, or, the veteran employee can offer the younger employee appropriate advice. As of August 2016, the ratio of people in the company's employ in Japan who are aged 60 or over reached more than 10%. The oldest employee is 84 years old while there has been a case where a certain technical advisor reached the age of 94.

Because Mayekawa Manufacturing Co., Ltd., realizes the significance of retaining employees aged 60 and over, the company operates its continued employment system. For seniors too, continued employment where they can make use of their experience is worth working for.

The British economist, John Maynard Keynes, wrote an essay in 1930 titled *Economic Possibilities for Our Grandchildren* in which he proposed that the economic problem may be solved within 100 years, or in other words, in 2030, and people would be freed from having to work. Even so, Keynes recommended that we work three-hour shifts per day. At the time of writing the essay, it seems that a sort nervous breakdown was common in England and the United States amongst the

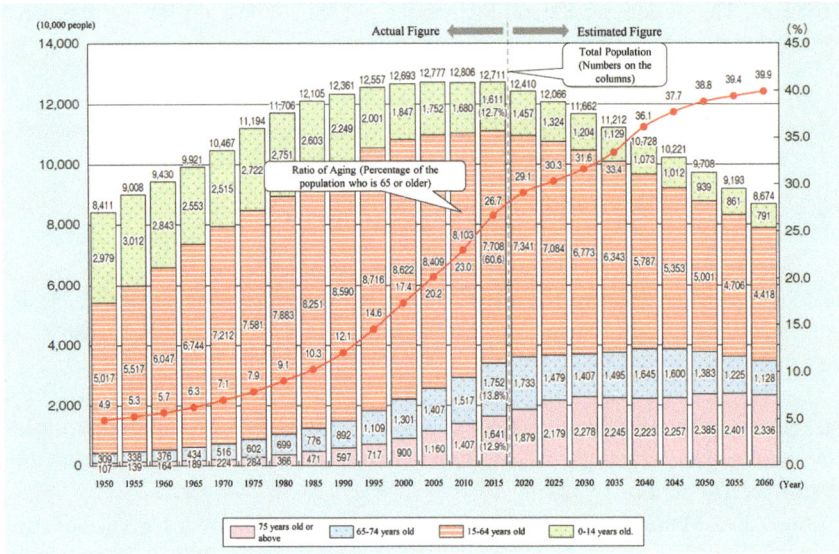

Fig. 6.9 Changes in aging and future projections
Source: "2016 Annual Report on the Aging Society," Cabinet Office
Sources: The "National Census" by the Ministry of Internal Affairs and Communications (until 2010), "Estimated Population (Presently confirmed number as of October 1, 2015 based on the 2015 preliminary reports of the population in the National Census)" by Ministry of Internal Affairs and Communications (2015), the estimated results calculated from the hypothetical median birth and death rates in "Japan's Estimated Population in the Future (calculated January 2012)" by the National Institute of Population and Social Security Research (from 2020 onwards)
Note: The total numbers from 1950–2010 include unknown ages. To calculate the ratio of aging, we subtracted unknown ages from the denominator.

wives of the well-to-do classes. Having a place in society through work that is valued and the feeling of being of help to someone is essential for maintaining one's mental and physical health.

According to the Cabinet Office's "Annual Report on the Aging Society: 2016," the percentage of the population aged 65 years or older was 26.7% in 2015. It takes 2.3 people who are now employed to support one elderly person. By 2050 the percentage of the population aged 65 years or older will rise to 38.8% and one elderly person will be supported 1.3 people who are employed (Fig. 6.9). Medical care, nursing care and age pensions will reach a deadlock if things continue this way.

To begin with, the idea of workers being aged from 15 to 64 was modelled on the industrial production in the years of Japan's economic bubble and is not in line with contemporary society. The percentage of today's youth, who are fewer in number than before, continuing on to university has risen so that half of them will not start work until they are 22 years old. On the other hand, there are many people who are aged 65 years or older who can still work hard.

If those active seniors can enjoy an independent life, that will lighten the load for the younger generation who are working, as well as ease worries about the nation's finances. Moreover, the knowledge and technical knowhow of active seniors is essential for increasing new industries that will support the Platinum Society. Hopes are pinned on the activities of seniors for education suited to the Platinum Society— not regular school education but practical education for younger employees, as conducted at the Mayekawa company. There should be many opportunities for this kind of activity. No matter how much mechanization and automation advance, master skills are still necessary in the industrial world. "Remaining active for one's entire life" improves the QOL of seniors as well as being a significant motto for the Platinum Society overall.

6.4.2   Making Use of the Knowledge and Experience of Seniors for the Next Generation

Trials are being made to make wide use of the knowledge and experience of active seniors in the field of education.

The city of Kashiwa in Chiba prefecture, the University of Tokyo's Institute of Gerontology, and the Urban Renaissance Agency (responsible for housing) are currently promoting urban design for a society of longevity. Nextph, is a learning space for children from Grade 3 in elementary school through to children in Grade 3 of junior high school, is operating as a partner business. Perhaps the most unique feature of Nextph is the Robot Club which has senior staff as its tutors.

Nextph has involved corporations, schools and local residents in its activities. Active seniors, who have already retired, teach English and math classes at Nextph. One of the senior tutors is Tatsuo Muto who was an engineer at Mitsubishi Chemical Corporation. Mr. Muto initially participated as an English teacher but after a while, he decided that he want to teach the children real science and proposed the robot class. After he began a class for which he developed original materials for teaching programming and so on, it filled up very quickly and has been extremely popular. Currently, the number of classes has been increased and a number of seniors have taken on the role of tutors.

With the advance of ICT, programming is already becoming mandatory in junior high schools while from 2020, it will also be a mandatory subject in elementary schools. From that social backdrop, there is a demand for programming classes and robot classes are well attended right across Japan. However, the number of people who can teach such classes is limited. People who are now employed are busy enough with their own work and do not have the leeway to participate in education. That is why participation by seniors is necessary.

Active seniors like Mr. Muto have a wealth of knowledge and experience they have accumulated during their time in the real world, and they can provide children with lessons taken from real life, different from classes at school or a cram school. At the same time, for the seniors, being able to participate in society of their own volition gives them meaning. For seniors to live their lives in good mental and physical health, being felt needed by society and continuing to be connected to society is important.

Therefore, the Platinum Society Network started up Platinum Mirai Schools. These schools are a near-future active learning platform where platinum master seniors and children full of curiosity learn about sustainable society building through exciting and top quality ways.

Various topics were considered for teaching at the schools but first of all, a robot class was begun. It is intended to enlist senior staff through collaboration with local municipal entities, employment centers for people with special talents or skills, or by using the alumni network of university teachers. Students at the Platinum Mirai Schools pay for their tuition monthly and that goes towards the remuneration of the staff. Through this method, the children attending the schools and their parents, and the staff members can tackle this initiative as proper work and by setting up a system where the money rotates, it is possible to manage the school in a sustainable manner instead of some kind of temporary activity.

The schools have been started up in Tokyo and Nagasaki. They will be used as a model to open further schools nationwide. Classes will expand to include not only robot classes but also English classes and environmental issues and so on, themes of a highly social nature.

6.5   Diverse Options and Freedom of Participation

6.5.1   Why Are Bonds Being Sought After Now?

Lifestyles and workstyles are becoming more and more diversified. This also includes diversification of the connections between society and individuals. As agricultural people, Japanese people have long been connected to land. Beginning with growing rice, work has had its roots in the local community, leaving no other option than to belong to the local community group. Anyone who disrupted that unity was ostracized and could lose their work as well as their domestic base. Not all that long ago, corporate culture was somewhat close to that. Lifetime employment was the base for life and participation in company events such as parties, sports days and company-sponsored recreational trips for employees were practically compulsory.

The local community that was bound together by rice-growing, and companies from some time ago, were places for forging bonds through group work. An affluent society was made possible because of the solidarity in such groups. However, now that people have acquired that affluence, its value has faded. These days, whenever something occurs, there is a clamor about the importance of bonds but that could be due to an instinctive feeling of crisis about the disappearance of such bonds.

Bonds are formed when people take part in group work earnestly. If so, then group work for creating the Platinum Society could take on that role too. These days, aside from working to earn one's livelihood, people often have some other kind of work they do because it gives their life meaning, or else they do something pro bono to contribute to society by using their work skills. Also, people are show-ing interest in participating in activities for communities that share specific values different to work, family life or hobbies.

With advances in technology, spatial distances have shrunk and many more options are available for working styles and life styles. Until now, people were more or less forced to participate in activities at the place of their affiliation such as com-panies, but from now on they are free to choose for themselves as to what kind of activities they do and what kind of communities they join. It is up to the individual to decide on what kind of community they will participate in, or indeed, whether they will participate or not.

6.5.2   Freedom of Mobility Induces Changes to Work Styles

In 1964, the year that the Olympic Games were held in Tokyo, the Tokaido shinkan-sen (bullet train) started running between Tokyo and Osaka, shrinking the travelling time to 4 h. The next year it became even quicker at 3 h and 10 min. Before the bul-let train started operations, this journey took 6.5 h so the new travelling time was a great reduction compared to before. Nowadays, it takes only 2.5 h. When the new Chuo Shinkansen, a maglev line, commences operations this journey will only take about 1 h. By air, it takes about 1 h from Haneda Airport in Tokyo to Itami Airport in Osaka. Shinaksen networks apart from the Tokaido line have been developed and the areas within reach of a one-day trip from Tokyo have definitely increased.

Innovations in transport will occur from now on too. Due to advances in ICT, the cost of transferring information is practically zero. Transport costs for human beings are undergoing a slashing of prices. With the entry of low cost carriers, the prices of air tickets have fallen to about half of what they had been previously. If a plane simi-lar to that of an advanced drone becomes real, then cost structure will be totally changed. Technically speaking, that is quite possible. The Airbus Group has already successfully completed its first test flight while unmanned combat drones are being used by the United States to carry out bombing strikes in Syria. Being unmanned does give some cause for concern but human beings also make mistakes. In any case, drones that are chosen are sure to have strict safety measures in place.

At the same time the degree of freedom in transport is expanding, the freedom not to move at all has also been created. Previously meeting up at a company and having face-to-face talks was the norm, as was reading documents that were circu-lated and stamping one's seal on them. However, nowadays most deskwork is done on computers. The system for holding video-conferences has become widespread and opportunities for everyone to gather together have noticeably decreased. Foreign-owned companies and major corporations have adopted telecommuting more and more. The number of people who do not have to get on trains at the same time every day is increasing.

If the need to go to the office decreases, then it is no longer necessary to live within commuting distance of the company. Nowadays we often hear of people who have built houses in Tochigi or Ibaraki prefecture and commute to their workplaces in Tokyo by the bullet train. But if telecommuting has begun in earnest, then it is not even necessary to live in the Kanto area. Company employees could live wherever they wanted to, such as in Hokkaido or Okinawa. Ordinarily they would work from their homes and go to the office by bullet train or plane when the need arises. The diversification of transport methods and transport networks, and the development of ICT have begun to change people's concept of moving around.

These changes will probably change systems like the number of days people go to work and the number of hours they work. We already have in place systems like flextime and a reduction in the number of working hours but these systems are not perfect. "I returned to my workplace after maternity leave under the system of shorter working hours, but if my child comes down with a fever and I have to arrive late and leave early, then I feel bad about this." Some people are working like this, afraid they are giving trouble to their colleagues. "It is difficult for me to go the office very day now that I am busy looking after my child but I would like to work a few hours each day from home." Some people have the will to work but due to problems with the system, they are unable to work. The ultimate ideal would be for parental leave not to even be necessary. To carry this out, it would be necessary to have an appropriate appraisal system and a fair salary structure system set in place. This is a highly significant topic for Japan where the population is decreasing.

Through technological advances, gaps in physical and spatial terms and in terms of time are being filled in, and working styles are being diversified. As a result of diverse options being proposed, the number of people who chose provincial cities may increase. For families with children, instead of having the children attend city schools that have hardly any space for sports grounds, it could be that children are happier racing around at full speed in schools in natural surroundings. Adults do not have to be exhausted from commuting to work on trains and they can enjoy their free time together with their children. These adults may choose to enjoy their own time in their own fulfilling way. If things went like that, then the overconcentration of people in Tokyo could be resolved.

Of course, there will always be people who prefer to live in cities. Being able to choose the lifestyle they like and the working style most appropriate for them is what is meaningful.

6.5.3 Spread of Multi-habitation

For the most part, we have succeeded in acquiring most of what humankind sought after in the twentieth century. People in developed countries do not live in want of housing, food, and clothing, and have been successful in evading the fear of dying due to a marked rise in longevity. Because of advances in means of transport and also in ICT, physical and spatial distances are narrowing more and more. People of today must decide on how to live in a world this free.

One answer is multi-habitation, or, the use of more than one residence. Nowadays, when mobility is much easier, a way of life wherein people live in a condominium during the week when they go to work, and then spend their weekends in a detached house in a provincial city, is a possibility. In the future, as autonomous cars become a reality, even people living in Hakone or the foothills of Mt. Fuji may commute to Tokyo. If they do not have to drive themselves, they can do work in the car, even if the commute takes 1–2 h.

A place for work and a place to live in – how should they be designed?

The choice made by Minoru Yamamoto, COO of Aratana, Inc., which is based in Miyazaki prefecture, is quite interesting. Aratana is an IT venture company that specializes in e-commerce websites and marketing tailored to such websites. Because of its line of business related to e-commerce, the company's location in Miyazaki prefecture has not posed any problems. Ninety percent of the company's clients are in Tokyo. Yamamoto himself is from Mie prefecture but he decided to move to Miyazaki so that he can continue his favorite hobby of surfing. "One step inside the office and it is the same business environment as Tokyo. One step outside and Miyazaki is a wonderful place to live in," says Yamamoto. He frequently enjoys surfing on his days off.

Past work options were split between either going to industrial areas on the Pacific Coast side of Japan, as symbolized by the employment en masse of middle or high school graduates from the rural districts during Japan's economic growth years, or to stay in one's local area and work in local industry. The first option meant leaving a natural environment but the latter option meant that the kinds of employment to be found were limited. However, now it is possible to live in a provincial city for the sake of one's hobby and yet be employed in the same way as if one lived in Tokyo, meaning one can enjoy the best of both.

As part of its management concept, Aratana has the ideal of employing one thousand people in Miyazaki and is actively promoting the employment of city-bred college graduates in the regional cities as well as those returning to their hometowns after having lived and worked in a big city.

6.5.4 Tokyo Work Styles and Countermeasures for Declining Birthrates

Not only venture companies like Aratana, but a trend seems to be emerging for major corporations to choose provincial cities too. In March 2016, YKK AP, manufacturer in nonferrous metals, transferred part of its headquarters to the city of Kurobe in Toyama prefecture. The YKK Group has an R&D base in Kurobe and by using natural resources like the wind, sun and underground water to the maximum, the company is promoting the development of its Passive Town Kurobe Model, a residential development which can keep electricity costs low. Known for its famous Kurobe Dam, this area is rich in water resources and abundant nature, making it a very comfortable environment to live in. With the opening of the Hokuriku bullet train line, access to Tokyo has become much better. The company intends to promote community-based development and use this in making its products.

By moving the company's headquarters and functions to a provincial city, the residential environment of the company's employees has undergone a major change. Rents are cheaper and with the shorter commute, employees are able to enjoy more leisure time. It is hoped that this excellent environment will lead to the promotion of health too.

Komatsu, a Japanese company that enjoys a major share of the world market in construction equipment, has proved there is a significant difference in the lifestyles of employees in Tokyo and in provincial cities.

Komatsu's headquarters are based in Tokyo but in 2001, the company set forth guidelines to move the headquarters' functions and partially they have made a move back to the city of Komatsu in Ishikawa prefecture, the birthplace of the company. The company's bases in Japan can be broadly divided into three groups: Tokyo, where the headquarters are; Osaka and the North Kanto area, where the production bases are located; and Ishikawa, where the company originates from. A survey on the average number of children per married female employee in each group revealed 0.9 for Tokyo, 1.3–1.5 for Osaka and the North Kanto area, and 1.9 for Ishikawa. The figure for Osaka and the North Kanto area is about the same as the national average but Tokyo and Ishikawa are clearly different. The percentage of female employees aged 30 years or older and who are married is 50% for Tokyo, 70% for Osaka and the North Kanto area, with Ishikawa taking the top ranking at 80%. Although the company system is the same, the percentage of female employees who are married is lower in Tokyo and also the number of children is lower. On the other hand, Ishikawa has a higher percentage of female employees who are married and also the number of children is higher. In general it is said to be difficult the women in executive positions to have children but that theory seems to only hold true for Tokyo.

Japan's total fertility rate in 2015 was 1.46. Although this represents a slight rise, it is not high enough to stem the decrease in population. The government devised measures to counter the declining birthrate but we are still nowhere near the target of a birth rate of 1.8. The Komatsu company's case may provide some hints for overcoming this problem.

6.6 New Industries Created in a Platinum Society

6.6.1 Marunouchi Platinum University – Thinking About Regional Issues in a Big City

If industries change and workstyles change, the knowledge and abilities required of individuals will also change. The kind of mature affluence to which the Platinum Society aspires is not something that is handed out. It must be seized, or at least found, by the individuals themselves. QOL improves because it is something the individual choses of their own will from diverse options. What is needed now is a place to study, where people can polish up their skills for the Platinum Society.

In July 2016, the Marunouchi Platinum University was opened in a business district in Tokyo. The career courses serve as a venue for lots of people to undertake new studies and to take up new challenges. Initially it was intended for business-people in their forties or fifties who work in the Marunouchi, Otemachi and Yurakucho areas. However, after it was launched we found that the classes were full of a diverse range of students, from those in their twenties to the senior generation who had already reached retirement.

The Town Revitalization by Outsiders Course began as soon as the university opened. Office workers from big cities have to think about revitalization of areas that are trying to deal with various problems. Over a period of 4 months the students learn about three cases and create business plans for each of them.

One of the cases is that of Isen town on Tokunoshima in the southwest island group of Kagoshima prefecture. What made this town, with its popular of just over 7000 people, so famous throughout Japan is the longevity of the island residents and the island's fertility rate. Of the Japanese people listed in the Guinness Book of Records for their longevity, two out of the three are from Tokunoshima. The number of centenarians on the island always numbers twenty or more while the average life span is more than 80 years of age. The total fertility rate is about twice that of the national average at 2.81. The island has boasted this top record for two consecutive years.

Mayor of Isen, Akira Okubo, who was a guest lecturer at the university, said that the reason for the town's high birth rate was that the local community was still quite strong and helped to bring up the local children. Tokunoshima is famous for bull fighting and it is still very popular in Isen. Mayor Okubo said that bullfighting gen-erates a lot of energy that excites people of all ages and both genders. This energy may be behind the longevity and the high birth rate of Isen (Fig. 6.10).

But even Isen is experiencing a decrease in population, and revitalizing the town and securing employment for the future are major challenges. At the Marunouchi Platinum University, such challenges are shared and people who are not from Isen but who are business people living in Tokyo, try to think up ideas for overcoming these challenges. The students at the university are given homework and as they are attending the course of their own volition, each person has a very serious attitude towards their study.

Tomoo Matsuda, Vice President of the Marunouchi Platinum University, said that usually meetings are carried out within the vertical superior-subordinate rela-tionship structure of Japanese companies but meetings at the Platinum University are held with students are of differing ages and with different kinds of work. So it is possible to have a discussion with everyone on the same level, thus causing an unex-pected chemical reaction. People from the financial world stress that sociality is vital while those from NPOs stress that if ideas are not feasible from the business side then they will not be sustainable. One of the appealing things about the univer-sity is the unpredictability which is surprising in itself.

Fig. 6.10 Mayor Akira Okubo of Isenchō on Tokunoshima Island. (Photo by Marunouchi Platinum University)

Not only Isen but also the cities of Miura in Kanagawa prefecture, and Hachimantai in Iwate prefecture are taken up as cases for study. Miura, famous for bluefin tuna, is a middle-ranking city which has the Healthy Peninsula Plan as its theme. Hachimantai is a famous highlands resort area with hot springs and skiing but it is in need of a business plan for revitalization of the region. The university's students study urban planning through these three examples of the isolated island model of Isen, the suburban model of Miura, and the highlands resort model of Hachimantai. Study is not limited to the classroom but students can go out and do fieldwork if they so desire. It is hoped that the students who have graduated from this course can become leaders in carrying out regional revitalization making use of the knowledge and experience they have gained in the course of their studies.

6.6.2   Developing Human Resources for Realizing the Platinum Society

The Platinum Society Network recognized from the time of its establishment that human resources who can be the driving force for regional development should be nurtured. Therefore, the Platinum Society School was set up. It is aimed at local government employees who are indispensable for overcoming local issues. Such people understand the situation and mechanisms. Because we want this kind of people to become future leaders, rather than young people or seniors, we get people from this group in the middle to take part.

At this school, students acquired practical knowledge such as cognitive power, ability to resolve issues, and leadership and management abilities which enable them to solve problems. New networks that bridge municipalities are formed. They must come to Tokyo 2 days a month but they share a learning environment with other students in similar circumstances and there are certain things they can only gain through an earnest exchange of opinions. It is hoped that the students can forge relationships wherein they can rely on their fellow students for help with just one phone call, 10 or 15 years later.

Over a period of 6 months, students compile all their work into "Platinum Concept for Our Town," a plan for implementing a project feasible for their own local municipality. This is not just brushing up some plan that already exists but is a proposal for what should be done anew as a solution in light of issues faced by the local community. Even if their proposal cannot be carried out right away, their experience of trying hard to come up with some concept will certainly be of use to community activities.

The Platinum Society School is now entering its ninth term of operation. Many graduates have commented that they wish to implement a Platinum Society School in their own local region and in 2014, The Platinum Society School @ Local Municipalities kicked off.

The basic way of thinking for this is the same as the Platinum Society School but the actual curriculum is customized to suit the location. For example, there was some news about a certain municipality wanting to educate its employees to become next-generation leaders. Therefore, a competition was held that focused on analyzing problems particular to that community and drafting proposals to overcome those problems. As doors should be open to local residents too, various programs were tailored to local communities such as a school that included the participation of residents, and also a school focusing on the topic of reconstruction after earthquake disasters was conducted. In the 2 years thus far, about ten municipalities have taken part including Toyota in Aichi prefecture, Higashi-Matsushima in Miyagi prefecture, Toride in Ibaraki prefecture and also in Saitama prefecture. As many enquiries are received, the number of schools conducted in the future will probably increase.

Human resources development by the Platinum Society Network is expanding across various age groups.

The Platinum Development of Future Human Assets is held in a training camp style during the summer holidays for up to 1 week at the longest (Fig. 6.11). Experiences learned from those of the same generation with diverse ways of thinking while sharing eating and sleeping arrangements together as well as having contact with such first rate teachers as Yasushi Akashi, former United Nations Under-Secretary-General for Humanitarian Affairs, and Dai Tamesue, former elite

Fig. 6.11 The project to foster future human resources @ Aizu. (Photo by the Platinum Society Network)

track and field athlete, is sure to have a major impact on the future of these impressionable young people. This initiative began in Aizu Wakamatsu in Fukushima prefecture and has also been conducted at Kashiwa in Chiba prefecture and Kikuchi in Kumamoto prefecture. University students have participated as tutors together with active seniors and this school has evolved as people from diverse age groups have built it up.

Various schools specializing in a particular field have become available too. A Platinum Energy School was held for junior and senior high school students to consider energy-saving and energy-creating in the town of Noheji in Aomori prefecture and also in Saitama prefecture. A Platinum Public Health Nurse School focusing on community health has begun in Fukuoka prefecture at Hisayama town which is the largest base for epidemiological research in Japan. In Tokyo, results of the fusion with administration can be observed.

The human resources development program has become quite substantial in terms of both quality and quantity. However, it had its start in just one platinum seed in the Platinum Society School. That seed travelled across different fields, generations, and communities as it put out new buds and roots and began to grow. There are still many platinum seeds lying dormant in the community. It would be ideal if the graduates of the Platinum Society human resources development program could find the seeds and help them to grow.

6.6.3   Education Changing Through ICT

Education is a new industry in the Platinum Society. Education different to what we have known thus far is sure to be created. The way that education should be may change dramatically due to the wave of internationalization and computerization. Education until now has been a place to gain as much knowledge as possible. However, even if enormous volumes of knowledge are gained, there is no hope to compete in a society where large amounts of information easily go to and fro across borders.

In the coming years, children will need to have the ability to create new values in a community of a global scale which will be made up of different cultures and sense of values. With the change from an industrial society to an information society, a twenty-first century education model for developing abilities will be sought after.

The computerization of education is indispensable. New information devices such as electronic blackboards and tablets, and digital education materials will soon be the norm. The importance of using ICT in the field of education will increase as materials made easy to understand with images and sound are developed, together with repetitive learning using computers. Teachers and students will be connected through the internet for teaching and learning together so that technically, no matter where you live in the world, you can always be in an educational environment. Also, classrooms will be freed up to the outside, making it possible for parents and local residents to participate in conversations while classes are being held.

The importance of computerization and digitalization has already been expounded upon at length but actually the field of education seems to be an exception, and with no progress being made in positive discussions. Especially in Japan, the demerits of computerization of education have been pointed out and it can be said that Japan lags behind other nations in the computerization of education. According to the Ministry of Internal Affairs and Communications (MIC), the number of students who use ICT both inside and outside of schools in Japan is particularly low making them noticeably behind other countries. Looking at the situation for ICT infrastructure in schools, according to research by the Ministry of Education, Culture, Sports, Science and Technology (MEXT), the number of computers used for education is limited to one computer per 6.2 students.

The Association of Digital Textbook and Teaching was established in July 2010 to resolve this situation. The purpose of the association's activities is to promote an environment where all elementary and junior high school students in Japan can study using digital textbooks. Specifically, this means (1) preparing 10 million tablets for use; (2) Developing digitalized versions of all the textbooks and learning materials; and (3) achieving a LAN penetration rate of 100% for Wi-Fi inside classrooms. The association is aiming to achieve these goals quickly, earlier than planned by the government. For that to happen, a three-party group must be formed consisting of the users (those at the schools, the students and the parents, and so on), the public service (the government and municipal entities); and the private sector (such as DiTT).

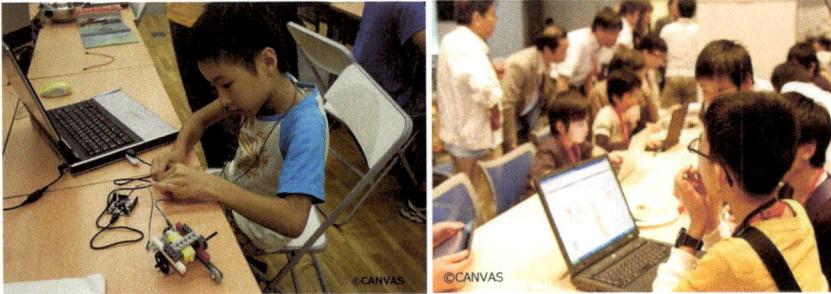

Fig. 6.12 Programming workshop conducted by the NPO, CANVAS

The situation seems to be changing at last. MEXT and MIC have worked together to conduct empirical research at schools across Japan. The debate about institution-alizing digital textbooks has begun in earnest. Provision of information devices and an internet environment is receiving a lot of support. Furthermore, based on the belief that the Fourth Industrial Revolution will occur, MEXT has announced mea-sures to adopt classes about programming from the stage of elementary schooling.

Local governments including those in Osaka, Arakawa ward in Tokyo, together with the cities of Takeo in Saga prefecture and Bizen in Okayama prefecture have given information devices to the elementary and junior high school students in their areas. Currently, there is a movement underway in 158 municipalities to adopt digi-tal textbooks. It is clear that if the heads of local government are eager about this initiative, it is definitely possible to implement.

Private enterprise is also making a serious effort in this regard. So far Microsoft, Intel and the NTT Group have been taking the lead in collaborating with schools and building up the number of precedents of the computerization of education. The number of corporations participating has increased with the services provided and areas covered becoming more diverse. The number of corporations introducing spe-cial tablets for correspondence courses is rapidly increasing.

Debate in the private sector to make programming a required subject in compul-sory education is becoming livelier. The NPO Canvas has had "Learning through Programming" not "Learning about Programming" as its slogan since 2002. Canvas has strived to develop curriculums, train instructors and organize a support system though collaboration with municipalities, boards of education and schools through-out Japan. In recent years, the number of learning centers teaching programming has surged. It is hoped that programming will be introduced in schools making use of the knowledge from the private sector (Fig. 6.12).

Changes arising from computerization are not restricted to elementary and sec-ondary education. Japan Massive Open Online Courses (JMOOC) has lectures by professors from such universities as the University of Tokyo, Kyoto University and Keio University available online for free. MOOCs were originally started up by Stanford University, Massachusetts Institute of Technology, Harvard University and the University of California, Berkeley. Anyone can get the world's best education for free as long as they have access to the internet. The number of options open to students has increased but viewed from a university perspective, it means they are compared to universities around the world. Some say that MOOCs could totally undermine the existing business model of universities.

Computerization and digitalization are making advances but there are still many problems to be addressed. The biggest problem is cost. Japan does not spend a lot of money on public education. When comparing costs of education in Japan to the country's GDP, Japan ranks lowest out of all OECD member countries. Japan needs to invest more in the computerization of its education system.

Another problem is that digital textbooks are not recognized as official text-books. Legally, textbooks are defined as books and must be in paper format to be officially recognized as a textbook. At long last, MEXT has begun to make revisions to the system but 4 years have been wasted since the author and others have first made such suggestions.

The computerization of education is not just an issue for national and local gov-ernments but is also an issue for all Japanese people, including those involved in education, and the parents of schoolchildren. Local municipalities, the government, the Diet and private enterprises each have a role to carry out. It is important to advance the development of both hard and soft infrastructure, support for teachers and a revision of the legal system concerning the computerization of education.

6.6.4   Adult Education as a New Industry

In 2013, Associate Professor Michael Osborne published a paper that stated that in 10 years' time, advances in AI will mean that many kinds of occupations will be taken over by computers. This sensational news was also reported in Japan. Various kinds of work ranging from jobs for entering data or being a supermarket cashier, which are mostly part-time, to occupations which require knowledge and experi-ence, such as accountants and loan officers, are supposed to be taken over by machines.

Not all of that may be true but undoubtedly, many kinds of work will disappear. Already a lot of jobs have been lost due to the introduction of IT. A typical example is factories that carry out mass production. Various processes have been automated and industrial machines can do the work with unrivalled accuracy, leaving no place for human beings.

On the other hand, it is true that due to the introduction of IT, work which didn't exist before has been created.

Construction is one area late to introduce IT. However, the Ministry of Land, Infrastructure, Transport and Tourism (MLIT), has launched i-Construction, a pro-gram aimed at increasing productivity in the entire construction production system and creating appealing construction sites. With this program, through the introduc-tion of IT, computers have taken over a large amount of work previously done by hand. However, not everything can be switched over. Some work, like maintenance, can only be done by human beings.

Much of Japan's social infrastructure is in a state of deterioration. There is nothing as difficult as the maintenance of facilities that have undergone repair work many times. As well as requiring excellent technical skills, the kinds of materials and work are not the same, so such maintenance has to be carried out on an individual basis which makes it extremely costly. Therefore, most infrastructure is left to deteriorate, leading to accidents such as the collapse of tunnel ceilings.

The introduction of a monitoring system, where sensors are installed in tunnels and on bridges, is now under consideration. Sensing technology and monitoring are two areas in which computers excel even though they cannot do the actual maintenance. The most appropriate kind of person for this work would be those who have worked on construction sites. If they receive special training and become IT engineers who can actually do the maintenance, then the problems of maintenance management of social infrastructure and of employment can both be resolved at the same time.

With the introduction of IT, the number of people working in the industry will no doubt be reduced. However, Henry Ford, who introduced mass production of the Ford Model T car more than 100 years ago, improved productivity ten times over and expanded the car market 100 times, and in turn improved profits ten times over. Even if we cannot expect such a bubble like that, if production efficiency at construction sites improves by ten times with the introduction of IT, then the existing market can expand while a new market including maintenance will occur and employment opportunities will increase.

"Learn a trade" is something often said but the kind of work one should take on will change according to the era. That change is happening faster and faster so people with their lives ahead of them need to always try learning something new. Society needs to develop educational institutions and educational systems to keep pace with those changes.

New business opportunities are possible regarding establishing educational institutions and qualification programs. New industrial education suited to a Platinum Society is needed. To that end, education and qualification programs themselves could become new industries.

6.6.5 Developing Leaders Who can Carve Out a Path to a New Era

In 2008, The University of Tokyo launched its Executive Management Program (Todai EMP). This is an educational program for those who will become future leaders and the program aims to develop human resources with high levels of general skills who can create a new era. Students are individuals in their forties with the potential to become top managers. No matter what situation they face, real leaders

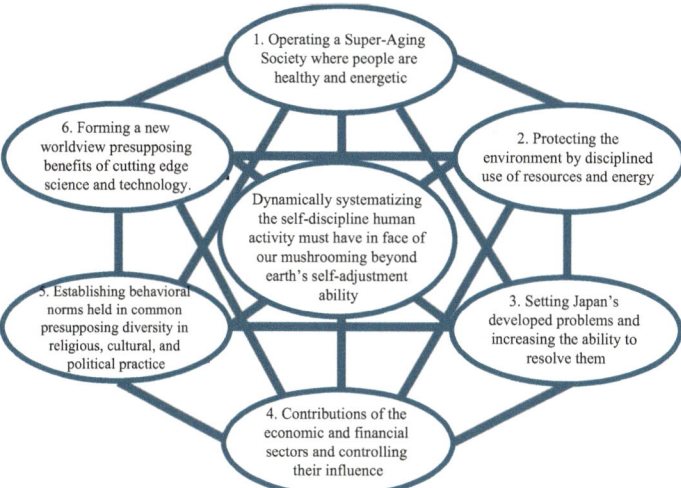

Fig. 6.13 Structure of The University of Tokyo's EMP "High culture and wisdom" program
The lectures cover a diverse field including: Religion, Philosophy, and Thought; Economics; Law and Political Science; The Global Community and Japan; Agricultural Science; Medicine and Health Sciences; Drug Development; Life Sciences; Neurology; Information Science; Media Studies; System Engineering; Material Engineering; Material and Energy Circulation and Environment; Space and Atoms; Architecture; Barrier-Free; Gerontology; Mathematics; History of Science; Cognitive Science, Social Psychology; Robot Engineering; Radiation; and, Innovation of Japanese Society.

must act based on solid knowledge and multifaceted ways of thought. Therefore, the program is designed so that students hold highly persuasive discussions about certain issues and they can propose and promote concrete measures to resolve those issues (Fig. 6.13).

Communication is divided into three areas and of the 180 classes that make up the program, 65% involve knowledge and insight.

Most of the classes are conducted by about one hundred lecturers from the University of Tokyo. With the broad range of knowledge the University of Tokyo has accumulated, opinions are exchanged during discussions while conveying ways of thinking to break ground with leading-edge knowledge. Rather than just learning about knowledge that is already complete, students try to understand the researcher's way of thinking, and the origin and background of how that knowledge was created, with the purpose of acquiring an even sharper insight. Attention is focused on nurturing skills for building logic that is universal and goes beyond culture.

Through a series of lectures, the structuration and integration of knowledge should advance. For example, in medical lectures the students can learn about possibilities for curing cancer and dementia, and students can build a common knowledge base. By imagining the impact this will have on society in the future, they can each deepen their thoughts on this by connecting it to religion, philosophy, thought and finance. If they conduct discussions after that, then they will probably gain an even deeper understanding and new awareness. Todai EMP is not a one-way educational program from lecturers to students but through lively discussion between the lecturers and students, or amongst the lecturers themselves, or the students themselves, everyone can learn from each other making it a place where new perspectives and opinions emerge.

In the lectures about space and particle physics, the students learn the importance of theoretical hypotheses, as well as the need to develop new observation equipment and machines. These days, even if advances are made in theoretical research and development technology, quite often simulations cannot be conducted, although this is not widely known. Regardless of a leader's field of expertise, they need to know that situations exist that can only be known about on the inside and also they need to know that problems do exist. Through these lectures, the students appreciate anew that in many fields, risk-taking by trial and error is necessary.

After the entire program has finished, the participants will have gained knowledge over a broad area and a multifaceted way of thinking. Graduates and the University of Tokyo lecturers hold meetings, research meetings, and on-site visits to different locations after the program finishes. Because each student is trying to advance their knowledge and participating in activities, the time spent in exchanges is fun and fulfilling. Graduates of the program number more than 320. Each year about 50 people are expected to graduate from the program, from now on. Todai EMP graduates are expected to be active in various areas of society.

6.6.6 Questioning Anew the Importance of Education

Since the Industrial Revolution, business activity globally has increased at a yearly growth rate of about 3%. In the future, with the growth of developing countries, this trend is sure to continue for the long term. Finally, in 2080, the average earnings per person globally will reach what it is in developed countries now.

However, there is no guarantee that current problems such as the economic disparity between nations and between individuals will move in the right direction and be resolved, or that regional conflicts around the world will be settled. Also, if humankind continues on with the current lifestyle that depends of fossil fuels, the effects of global warming will further accelerate the disparity between nations.

Society is continually changing and the increasing rate of globalization gives impetus to social changes. Globalization is no longer restricted to the economic sector. Various areas such as politics, science and technology, culture and so on, are linked to one another in their development. This is not to say no to new developments themselves. However, as a large volume of technology keeps being implemented into society at an even faster pace than before, social changes accelerate and become more complicated. Because of that, many people keenly feel how difficult it has become to show leadership.

Nonetheless, if the world's third largest economic power, Japan, is to continue developing, it is vital that those in a position of leadership understand the newest knowledge in various fields and how such knowledge is undergoing change. Then, by strengthening their ability to foresee the future and their ability to communicate, it is necessary to strengthen the impact on other countries.

Learning is not just necessary for those in leadership positions. Through advances with AI, computers, sensors, robots, and 3D printers, part of the work currently done by human beings will be taken over by these inventions. That does not mean that people's work will be taken but rather, new kinds of jobs and new work will also be created. At such a time, conventional knowledge and skills will not be enough to deal with this new situation. New knowledge and skills are necessary for tackling new work. Until now, it was possible to learn on the job and accumulate experience through work after one had finished school and become a working adult. But from now on, in order to keep step with the social changes happening, knowledge and technology must constantly be updated. A diverse educational system that supports people's learning is needed.

In the future, labor productivity will increase through the application of new systems and people will have plenty of time and opportunities to receive education after they have become working adults. Many people will receive education appropriate for the new times and partake in work that is advanced and of a high level, thus contributing to the further development of society.

6.7   The Platinum Society Becomes More Visible

6.7.1   How to Promote a Platinum Society

Even in Japan, which is a leading country in resolving societal problems, problems have come to light in remote island regions, making them problem-saddled regions.

Tanegashima in Kagoshima prefecture has a population of about 33,000 and a surface area of 445 square kilometers. Tanegashima is a narrow yet long island stretching from north to south. It is divided into one city, Nishinoomote, and the two towns of Nakatane and Minamitane. Out of all the outlying islands not connected to any of the four main islands of Japan, Tanegashima is the fifth largest. The rate of its declining population and the aging of its population are higher than the national average thus truly making it a problem-saddled region.

In 2012, the University of Tokyo's Organization for Interdisciplinary Research Projects set up the Presidential Endowed Chair for Platinum Society. Collaborations between industry, academia and the public, based on this Presidential Endowed Chair have been advanced, and this movement has expanded to a point where a total of nine universities are conducting activities on Tanegashima (as of August 2016). This provides university researchers with the opportunity to verify technology they developed themselves and brush it up as technology that can further enrich the local community (Fig. 6.14).

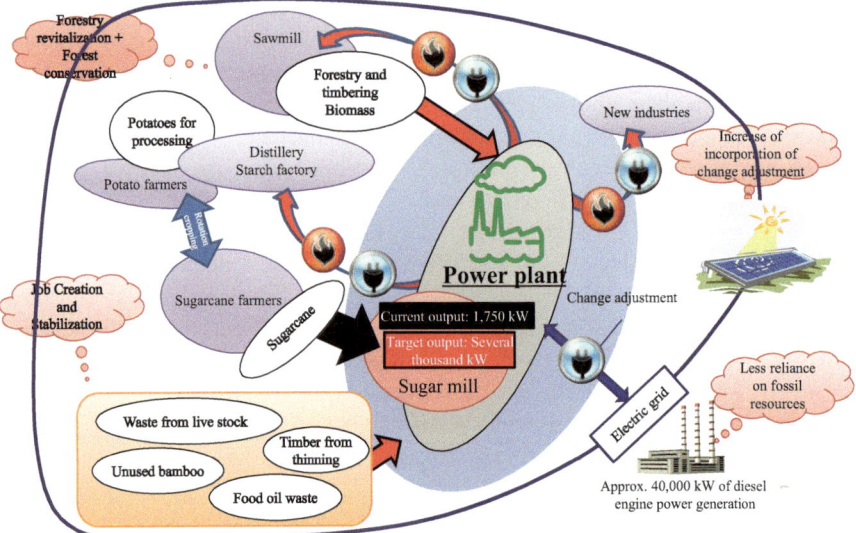

Fig. 6.14 Conceptual diagram of the agricultural and forestry collaborative system implemented in Tanegashima. Kikuchi Y. et al. *Journal of Industrial Ecology DOI*: 10.1111/jiec.12347 (2016)

Tanegashima has various kinds of agriculture and livestock industries such as sugar cane, sweet potatoes including the Anno variety for the fruit and vegetable market as well as other kinds of sweet potato for making Japanese *shochu* (white liquor) or starch, wet-rice cultivation, and the raising of Wagyu cattle. Out of these, sugar cane holds a special place as part of the island's culture and the sugar cane industry, including the manufacturing of raw sugar, is one of the key industries of the island.

Sugar cane juice is extracted by crushing the sugar cane. This juice is then clarified, made more concentrate and crystallized to make raw sugar. Fibrous solids that remain after this processing are called bagasse and from long ago, have been use as fuel inside the sugar processing factory. All the electricity and heat needed for processing sugar can be provided by the bagasse but research has shown that in most sugar mills, energy derived from bagasse is affordable and can become excess by increasing energy efficiency at sugar mills.

On the other hand, other vital island industries, the starch factory and farmers growing mangoes continue to burn fossil fuels to maintain temperatures in hothouses or for generating steam from water. This is a general-purpose system used in households too but from an engineering perspective, it is a waste to use fossil fuels in order to obtain temperatures lower than 150 degrees. If the leftover heat generated at the sugar factory could be used, it would easily meet the demand.

The system where different industrial facilities share materials or energy is called industrial symbiosis. This system is ideal but actually implementing this in the community is not as easy as it may sound. As well as technology for transporting heat, sorting out the connection with relative laws and creating a system for cooperation within the municipality, it needs to be clarified as to whether it is attractive for the local residents or not. This is a difficult role to be carried out by corporations and business operators alone. In that regard, the university has a neutral position in the local community. An effective method would be for the university to take on the main role and explain the academic significance from many sides, and for industry, academia and the public to find some way to grapple this.

On Tanegashima, several projects like this have been set up. In regard to sugar cane, there is a verification plant, the first of its kind in the world, for the inversion production process for sugar and ethanol. It received the Grand Prize for the Global Environment Award in 2013. In this case too, the university has conducted a systematic analysis for reducing the burden on the environment and for the adoption of technology. The method Tohoku University has for producing biodiesel using ion-exchange resin means that high quality biodiesel can be produced continuously from low grade waste cooking oil by the easy procedure of simply passing it through a resin packed tower. Facilities for verifying this technology on a scale for adoption have been built on Tanegashima, the first of its kind in the world. Trial runs using a common rail diesel engine are being conducted.

Researchers from Kyoto University, the Nara Institute of Science and Technology, and Kumamoto University are collecting unusual data from all around the world related to health management, by using a wearable sensor to measure changes in a person's heart beat to try to anticipate circumstances that could impact on the functions of the person's autonomic nervous system due to heat stroke and so on, and develop a mechanism to sound an alarm. These projects are also being used in education in the local community.

Lecturers from the University of Tokyo, Tohoku University, and Kobe University introduce projects at the local prefectural Tanegashima senior high school as well as getting the students to quantitatively specify local problems using the Regional Economy Society Analyzing System (RESAS). The students then conducted a project to consider ways to resolve those problems. The outcomes were suggested and transmitted at a symposium attended by the head of local government, industrial associations and local residents. Opinions were exchanged with seniors from public organizations and new knowledge was conveyed with a kind of fusion starting to occur. The appeal that local communities that have started to move towards a Platinum Society have is also spreading to further than university researchers and corporations hoping to develop business. In the Hands-on Activity Programs conducted by the University of Tokyo for undergraduate students, a program related to agriculture, forestry and fishing industries in Tanegashima and regional revitalization, was proposed for FY2016. Although this was proposed for the first time when the program was in its fifth year, out of the total number of 443 programs, the Tanegashima program ranked first domestically while it was ranked high by program applicants, including for programs conducted overseas. This revealed how students from urban areas were attracted to experiences in regions undergoing innovations. In this way, exchanges of new knowledge are being created one after another, like chain reactions.

6.7.2   The Platinum Network Society and the Platinum Vision Award

The Platinum Society is a model for growth in a mature society. A mature richness in terms of quality and a vision that seeks QOL is not confined to certain regions or countries, cities or countryside, or outlying islands. However, specific ways to achieve such a vision or initiative differ with each area. In looking at areas across Japan, there are many examples to be found that exhibit various kinds of resourcefulness and clever ideas. Areas experiencing certain problems may be able to get some hints by looking at how other areas with a similar problem handled it. By exchanging each other's ideas and knowledge, it may be possible to come up with something even better. Joining hands with each other will lead to a bigger movement giving them the power to demand legal and social reforms.

From this kind of idea, the Platinum Network Society was launched in August 2010 by its 46 founders. Even if social reforms are known to be necessary, all of society moving together as one is not a reality. So, first of all, the frontrunners need to go ahead and increase the number of supporters. Then, when critical mass is reached, everyone can move together. The people gathered here are the frontrunners who can make a move before others.

More than 6 years have passed since the Society was launched. The Society has grown into a large organization and as of September 2016, the Society has 84 corporate members, 154 local government members, 56 special members, and 6 members from outside of Japan, making a total of 300 altogether. As mentioned earlier, the Platinum Network Society conducts various activities such as developing various educational programs for fostering human resources, publishing the Platinum Society Handbook, holding symposiums, establishing individual working groups for health management, and holding discussion meetings to generate ideas (Fig. 6.15).

Amongst such undertakings, the Platinum Vision Award, inaugurated in 2013, has been often mentioned in the media and is credited with making Platinum Society ideals better known. Recipients of the Platinum Vision Award include local governments, corporations or organizations that have created a new industry through some kind of innovation, or are aiming to resolve regional problems by ingenious measures. Through their efforts, these governments, corporations, and organizations represent a society that is aiming to become a platinum society.

6.7.3   Creating the Platinum Society Handbook

The Handbook was created so that methods from these advanced examples can be used in other similar cases. Japan's rapid economic growth during the period between 1955 to the 1970s was maintained by the development of factories in regional areas as society became more industrialized. At that time, the development

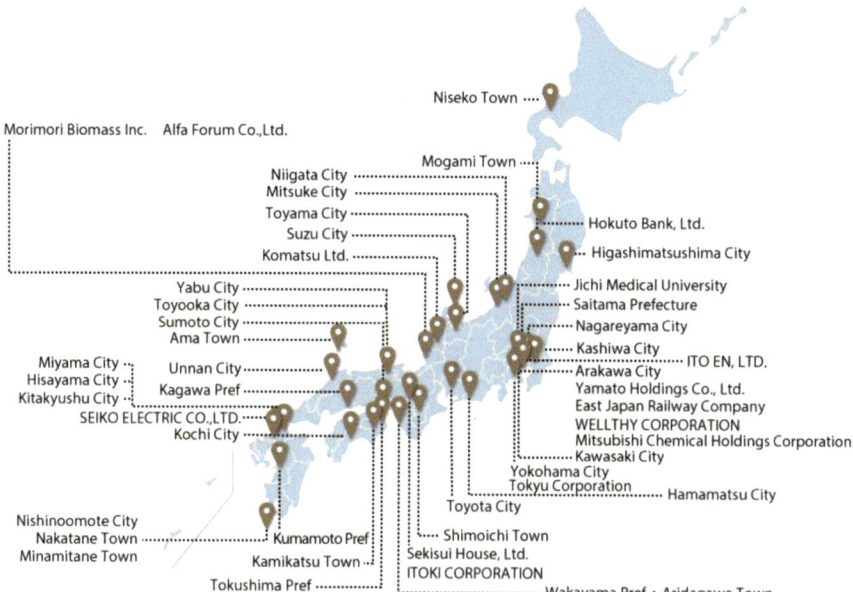

Fig. 6.15 Past organizations that received the Platinum Grand Prize

of factories in regional areas, and methodology for rolling out the same measures in similar cases were researched and retained in the form of textbooks, handbooks and manuals, and this supported the industrialization era. In order for a Platinum Society to evolve, we need to create methods for utilizing similar cases such as those that supported the era of industrialization. Figure 6.16 shows the overall image of structural analyses as listed in the Handbook.

The following content has been compiled and recorded in the Platinum Society Handbook.

1. Finding problems (regional) and setting goals
2. Projects that have been conducted (specific details, schedules, costs, systems, etc.) and factors for their success.
3. Local resources that were used.
4. Mechanisms and systems
5. Outcomes and future developments

Finding problems and setting goals should be kept as simple and clear as possible when planning and implementing specific projects. By keeping these simple and clear, the project is more likely to be successful. Specific details about the system for promotion and also factors for success or factors that inhibited past projects are included. These specific details include schedules, (the project's processes), costs, and bringing in experts as leaders. Numbers 3 and 4 are prerequisites for carrying out projects. Number 3 includes a review and rediscovery of materials unique to that area. Number 4 describes regulations as well as any subsidies which may be avail-

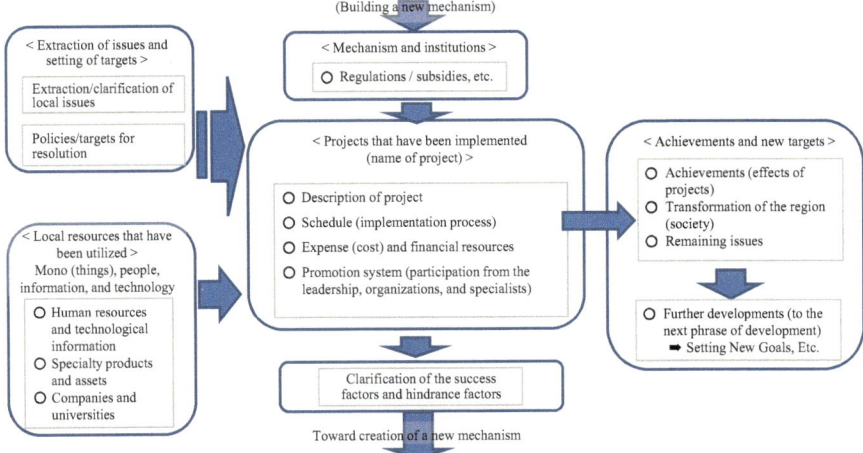

Fig. 6.16 Methods of creating a Platinum Society—The structure of a platinum vision handbook. Komiyama H, Matsushima K. (2012) Platinum Vision Handbook —Change the world through the power of active elderly. Platinum Vision Committee, Japan

able. Nowadays, examples that aim at promoting deregulation by utilizing the Comprehensive Special Zone System are increasing. Due to outcomes from this, further deregulation or new support measures are coming into force.

For development in other areas the results, including any impact the project had on the local community, are expressed in numbers so that the accuracy and details can be adequately grasped. Further developments are also described as it is possible that the local community may be impacted even further as the project continues to expand.

The basic intention of the entire Handbook is to make Input-Output relations as clear as possible so that cause-and-effect relationships are clarified. To this end, the IDEF system of analysis was used as a base.

Rather than examples, the Handbook focuses more on factors for success and the development processes of various cases.

For example, the senior high school in the town of Ama was in danger of becoming closed down. However, the implementation of a program to have students come to the island and study at the school successfully increased the number of classes. This excellent result was brought about by the sense of crisis shared by the town mayor and island residents that keeping the school viable was directly linked to keeping the island viable. As well as the entire town getting involved in this project, factors for its success are that experts were brought in from outside and that the project was carried out with collaboration and division of roles with these experts. Out of these, ideas unique to the island were born. They included conducting social education with the town residents as teachers, and opening a publicly run learning center.

Toyooka city has made extensive use of white storks and developed a special brand too. In order to raise white storks by artificial incubation and eventually release them back into the wild, the city had to undergo a major change in developing and expanding farming land that either used very little or no agricultural chemicals. The brand of organically grown rice known as "Rice that Raises White Storks" was produced and marketed. Now, efforts are being put into attracting businesses to further promote the economic environment. The city is being developed as an environmental and economic city.

The number of cases described in the Platinum Society Handbook is more than 30. Case examples will continue to be gathered and analyzed. As well as providing examples from diverse areas for reference, hopefully it will lead to the implementation of projects in many different locations.

From before, I have though that the ideal for various organizations is a distributed cooperative autonomous system. For example, in a human body, organs such as the heart and the liver function autonomously but in its entirety, the system is created to be one life. In the management of organizations, although the individual elements are active separately, it is better for the entire system to be properly balanced, as necessary. The Platinum Society Network is also based on this idea.

The Oki-Dozen Senior High School Appeal Project in Ama, the Zero Waste policy in Kamikatsu, the Miracle of Yanedan village, and the Platinum Triangle initiative centered around Futakotamagawa—what each of these locations is aiming for is quite diverse. These diverse communities each exist independently of each other but they are organically connected against a backdrop of information and freedom of mobility. That distributed cooperative autonomous system is what the Japanese Platinum Society looks like. Expanded to world-sized scale, it is the kind of world we hope to achieve in the twenty-first century.

6.8 The Platinum Society and Vision 2050

Vision 2050, which was promoted at the end of the twentieth century, is a macro vision derived from thinking sustainably about the Earth from the perspective of materials and energy. At this time, lifestyles and social systems could not all be included. However, in the 2000s, a vision was shown wherein Japan, as a problem-saddled developed country, should aim to become a leading country in resolving societal problems instead. A clearer image of this is the Platinum Society (Fig. 6.17).

As has been touched upon earlier, the Platinum Society is already beginning to sprout up in various locations around Japan. It is becoming accepted as a practical vision and not simply some armchair theory. It goes without saying that the Platinum Society will not be a reality if the Earth ceases to exist. A platinum kind of perspective is essential in regard to measures for ensuring Earth's continued existence. In other words, the series of visions that started with Vision 2050 are intertwined with each other and combining them into one vision paints a much bigger picture of the world.

The New Vision 2050, the main theme of this book, was born from such ideas as: each individual pursues a Quality of Life; regions are each attractive in their own

Fig. 6.17 Platinum society. (Source: The platinum society network)

unique way; society is sustainable without having to abandon such matters as economic development and reducing carbon. The New Vision 2050 logically shows that such things are a possibility.

The road is not smooth. Obstacles must be overcome and there are problems that have to be settled. At times it will be necessary to make mid-course adjustment. But even so, this is the road that must be taken. The reason is because the future for the world beyond the New Vision 2050 will definitely be bright.

Bibliography

Cabinet Office, Japan (2016) 2016 Annual Report on the Aging Society (JP only)

Kikuchi Y et al (2016) Industrial symbiosis centered on a regional cogeneration power plant utilizing available local resources: a case study of Tanegashima. J Ind Ecol 20:276–288. https://doi.org/10.1111/jiec.12347

Komiyama H, Matsushima K. (2012) Platinum Vision Handbook —Change the world through the power of active elderly. Platinum Vision Committee, Japan

Interview

Interview 1: Toyota Environmental Challenge 2050

Challenges Unachievable Following the Lines Laid Before

Takeshi Uchiyamada
Chairman, Board of Directors
Toyota Motor Corporation

Komiyama What has the response been like in-house and in the world at large to Toyota Environmental Challenge 2050?

Uchiyamada The goals we set for it were quite bold, and we needed to be fairly decisive about announcing numerical targets. Naturally, people in-house were surprised, and so were the parts manufacturers with whom we do business. "Why is Toyota talking like this," they wondered. On the other hand, we also quickly received proposals suited to the direction in which Toyota is headed. Furthermore, most of the companies in the Toyota Group subsequently similarly announced their own goals.

Komiyama I see.

Uchiyamada It was like this when we released the Prius in 1997, too. Even though we had come out with a car with good fuel economy that brought down the volume of CO_2 emissions, we couldn't get big results with the Prius alone. We got our customers to recognize that reducing CO_2 is one of the issues that a car-oriented society faces. As a result, since then environmental performance has come to be a factor upon which product appeal hinges and all car manufacturers compete on it. It goes without saying in the case of hybrids, but fuel economy has also improved with standard gasoline and diesel-powered vehicles. The volume of CO_2 emissions has fallen substantially. What's crucial, I think, is for a development that closely resembles this, where Toyota is not just saying something on its own but rather is creating a group of peers in the process who are all pointed at the same goal.

Takeshi Uchiyamada, Chairman, Board of Directors, Toyota Motor Corporation

1969   Joined Toyota Motor Corporation after graduation from Nagoya University in Applied physics
1996   Chief engineer of Vehicle Development Center 2 which developed the Prius
1998   Board of Directors
2000   Chief officer of Vehicle Development Center 2
2001   Managing director and chief officer of the Overseas Customer Service Operations Center
2003   Senior managing director and chief officer of the Vehicle Engineering Group
2004   Chief officer of the Production Control & Logistics Group
2005   Executive vice president
2012   Vice chairman of the board of directors
2013   Chairman

The Impact of Climate Change

Komiyama Please explain for us the three Zero CO_2 challenges and three Plus challenges you tout in Environmental Challenge 2050.

Uchiyamada The Zero challenges begin with achieving zero CO_2 emissions for new cards. To be precise, it calls for a 90% reduction of CO_2 emissions by 2050 compared to 2010. The next one sets the goal of reducing to zero the CO_2 emissions throughout a vehicle's lifecycle from production to disposal. The third challenge will have a particularly big impact for us and that is to reduce to zero the CO_2 emissions from vehicle manufacturing. This is all an attempt to manufacture vehicles with zero CO_2 emissions, and bring them close to zero when driving as well.

As to the Plus challenges, they show we are not only aiming at a zero-emissions world. They involve factors that may not be minuses even today, but are still attempts to create a better human society. One of these is water. We will reduce the amount of water we use at our plants, and clean the wastewater thoroughly. The next one is our effort to recycle by recovering materials for cars from cars that are slated for disposal. The final Plus is an initiative to further broaden the scope of activities to protect an environment including tree planting and taking biodiversity into account.

We are treating these initiatives as single package and are pushing to achieve them with the target of 2050 in mind. Of the Zero challenges, the goal of manufacturing cars with zero CO_2 emissions is one that cannot be achieved without some sort of breakthrough. We are aware that by creating challenging targets we will not be able to achieve them by following the lines that have been laid before. I think that a variety of ideas might come out of this.

Komiyama I'm delighted about the Environmental Challenge 2050 myself. The fact that Toyota made this declaration. The integration of its elements is wonderful. It's purely coincidental, but the thinking behind the Zero challenges closely resembles the Vision 2050 that I proposed in my 1999 book Chikyu jizoku no gijutsu. The Platinum Society concept I'm working on now likewise speaks to the three Plusses. How was the overall declaration put together?

Uchiyamada I think it was the same case with your Platinum Society idea, but I think we arrived at the same place when we thought about the way things ought to be.

Komiyama Was it discussed in-house?

Uchiyamada The discussions took place mainly in the Environmental Affairs Division. At the core of our vision was the desire to bring good products out into the world, create even better cars, and bring smiles to the faces of our customers. Based on this, we boosted our management foundations and create new products. It's an old turn of phrase, but one of the Five Main Principles of Toyoda called for contributing to the common good. It means being of use to the world through the medium of the automobile industry. This way of thinking is deeply imbued in our workforce. We don't think we are making cars for profits, but rather we want to be of use to the common good by making cars. Around the time we joined the company, the important thing was developing and deliver cars that were highly reliable and offered good cost performance so that everyone could drive a car, and through that make life more convenient. After that, we worked on building the Prius and making fuel-cell powered vehicles in our effort to be of use to everyone through cars that offer excellent environmental performance.

At that point, when we thought about what idea should be our next focal point, the one that we of course felt really struck home was the impact of climate change. The Intergovernmental Panel on Climate Change (IPCC) issued a report on the need to reduce to zero or even shift into the negative range greenhouse gas emissions by the end of the twenty-first century. However, this simply is not something that can be handled right away. That's why we set as our target year 2050, which is 50 years before the century's end.

Komiyama I'm sure you had some people in-house who thought this was impossible, no?

"Let's Do What We Have to Do."

Uchiyamada The debates we had were of whether or not to include the two numerical targets: "90% reduction in volume of CO_2 emissions for new vehicles" and "plant zero CO_2 emissions."

Komiyama I understand you gave people a push by saying, "This is not a matter of 'let's do what we can do,' but rather 'let's do what we have to do.'"

Uchiyamada That's something I say all the time, but this time I was really excited about it. I thought, "wouldn't it be wonderful if we really could make cars without producing any CO_2?"

Even when we developed our hybrid and fuel-cell powered vehicles, there was talk at the start about how it just couldn't be done. However, it's also a fact that once we set a deadline and went about it wholeheartedly we accomplished our goals. This of course is not something we are going to manage right away, but I think that by setting our goals and giving it go we should be able to do it. I don't think the reactions in-house have been all that negative.

Komiyama They call this kind of thing "backcasting." You have a goal that says this how things will be in the future, and then you decide what kind of technology to bring to bear.

Uchiyamada Yes, that's right.

Komiyama For techies, that kind of thing is a big incentive, isn't it?

Uchiyamada Yes, that's right. However, while it's fine to set goals, everyone has the awareness that the road to be travelled is going to be a fairly rough one. For example, the Prius did not take off in the marketplace right away, and what's more when we released it the business was in the red. If we had kept pushing it under those conditions, the red ink would balloon. It might be a bit overdramatic to say this, but it would have been easy for the company to have collapsed on account of the Prius. To popularize the model, we had to set the costs so we could generate profits with a hybrid. Given that the costs did not come far enough just with mass production, we completely rebuilt the Prius' hybrid system when the next model change came up. The cost of the second generation's system was half that of the first, while the third generation's was one-third.

Komiyama That's one of the effects of mass production, isn't it?

Uchiyamada I think changing the design was a bigger factor than mass production.

Komiyama That's unexpected.

Parts Manufacturers Have a Major Presence

Uchiyamada Environmental Challenge 2050 is the same. Given that we aren't necessarily going to get the capital from somewhere, we need to keep earning money while we also try to hit our targets.

On this point, I think the question of whether or not parts manufacturers are going to compete over research and development plays an extremely big role. All of the main components for the first-generation Prius were developed and manufactured by Toyota. However, once parts manufacturers come to think of this as a new business opportunity, the total number of engineers involved increases dramatically. Those people all wind up competing for us. In that sense, we fully recognize the importance of our stakeholders.

The Environmental Challenge 2050 is not a matter of Toyota arbitrarily coming up with targets that we can achieve through our own efforts. Collaborating with equipment makers, parts manufacturers, and distributors among other is indispensable. Furthermore, in the years ahead energy companies will also be part of it. To reduce the CO_2 emissions of new vehicles to zero, we naturally are keeping an eye on developments with fuel-cell powered vehicles, for example, as well as would-be "hydrogen society" The questions of whether Japan as a country can make hydrogen part of its core infrastructure and whether the energy industry will wade into the field are extremely important. Nobody is going to build a hydrogen-fueling station if you ask them to do it just because Toyota has built a fuel-cell powered vehicle. We are doing this because once everyone thinks that hydrogen is going to become an important part of the infrastructure we would then decide to put out vehicles that use hydrogen.

Komiyama Zero at the plants, too, is a difficult numerical target.

Uchiyamada The biggest problem when it comes to manufacturing automobiles with zero CO_2 emissions is the furnaces. Drying furnaces for paint and so on.

Komiyama Do you have that many furnaces?

Uchiyamada There's also the forging process. We use furnaces when heating and then drying materials. Our battery plants also need drying furnaces.

Komiyama At what temperature do they dry things?

Uchiyamada For forging, it's close to 1000 °C. The temperatures of the drying furnaces for painting don't get that high. Whatever the case, we're all thinking about how to short the total length of the furnace in order to save energy. Recently, when forging axle-shaped parts, we've been using high-frequency heating rather then send them through the furnace. We use electricity to apply heat. This greatly reduces the amount of energy consumed.

Komiyama Gradually switching to electricity will make it easier to use renewable energy.

Uchiyamada That's why I'm not gloomy about advances in technology.

Mid-To-Long-Term Targets for Clearing Regulations

Komiyama Did you announce Environmental Challenge 2050 last October with the 2015 United Nations Climate Change Conference (COP21) held at the end of last year in mind?

Uchiyamada It was that kind of a year.

Komiyama Did you think an agreement could be reached?

Uchiyamada Given that there no longer is any "later," I thought that—setting aside what level it would be—some sort of agreement would be reached. That's because a variety of organizations were also applying pressure along with that coming from governments.

Komiyama I was saying they definitely would make an agreement. That is to say, CO_2 emissions from the world's two biggest emitters, the United States and China, have already started to fall. These two countries were positive.

Uchiyamada The volume of Japan's CO_2 emissions is about 4% of the world's total. Even if Japan reached zero, the world's level wouldn't get that much better. As you say, the important thing is for countries like China and the United States, which are major emitters to participate in a global framework and worked toward reducing emissions. In that context, I think the question is what can an advanced country in the area of environmental technology like Japan do to help.

Komiyama If we don't get involved, the environment will do an about-face in the wrong direction. I have done an investigation and analysis of automobile catalogues using data from around 1995. What I found was that for automobiles of the same weight, the average fuel economy values for Japanese cars were about 15% better than those from the US and European manufacturers. However, if we look at data from 2015, we now see that there is practically no difference when it comes to standard gasoline- and diesel-powered vehicles.

Uchiyamada I talked about it earlier, but one of the biggest drivers for hybrid vehicles was that the world's manufacturers got wrapped up in a competition over fuel economy. Due to this, governments also decided there were still more regulations they could set and those regulations got tighter. It's tough for us, too, but given that these regulations apply to all manufacturers your only choice is to do what you can to come out on top under those conditions.

Komiyama Right. If everyone really joins in, and in particular if China and the United States join in, then the tighter the regulations the more Japan has an edge.

Uchiyamada There certainly are some who say it's better to make them as strict as possible, but when you do so the burdens on the customer rise dramatically. You need time in order to not push the cost up. The best time for developing and introducing the new engines and platforms that make the biggest improvements to fuel economy is when you do a model change. Accordingly, even people say let's improve it by 20% starting next year you can't do it out of the blue. Mid-to-long-term targets like those of the Paris Agreement talked about at COP 21 are important.

Komiyama Are various efforts already underway with 2050 in mind?

Uchiyamada Things have already started. I can't offer you any specifics as of yet, but let's take furnaces as an example. Currently, we fuel them with natural gas, but we're wondering about whether or not we can use hydrogen, which doesn't produce CO_2 when burned. Nobody thought about this at Toyota before we came up with the challenges. It's that sort of a new idea. We haven't built a plant in the past 3 years, but in the future, we will begin construction for the next stage of development. Because we must test technologies that have possibilities, we are studying a variety of things like whether it will now be possible to try taking those ideas on board. The hurdles we have to traverse are high when it comes to something like making a zero CO_2 car by 2050 since that will be done not just at one plant but at plants around the world.

Komiyama The effects will be enormous, so I really do hope you will put in the effort.

Uchiyamada Even so, the things that one company can do are small compared to the problems the world faces. As you said earlier, if everyone makes an effort, it will create great energy for dealing with the problem of climate change.

Interview 2: Regional Revitalization and New Work Styles

Local Activation for Stronger Competitiveness

Masahiro Sakane
Councilor, Komatsu Ltd

Komiyama There are three topics I'd like to cover today: The transfer of some Komatsu headquarters functions away from Tokyo, ideas for implementing different work styles in the company, and the kinds of steps that will be needed to achieve regional regeneration in Japan as a whole.

Sakane I can talk about several things. Before the start of the Abe administration, when the LDP was out of power, they said "Talk to us when we set up the Headquarters for Japan's Economic Revitalization," which then got started inside the LDP offices. It was at the first meeting, I believe. What I said was, the two big challenges for Japan are overcoming deflation and revitalizing the regions, and when it comes to deflation and regional decline, Komatsu as a company is a microcosm of the country.

Let's start with how Komatsu is a microcosm for deflation. You know as well as anyone about our "SMARTCONSTRUCTION" business, which brings new solutions to the worksite by using the internet of things (IoT). The original motivation for that had to do with the fact that Japan is the least profitable place in the world to sell construction machinery. We do sell a lot of construction machinery in Japan, considering the country's small land area. And that's because of the organization of the construction industry, which extends to fourth-tier and fifth-tier subcontractors, meaning there are a great many companies. They all lease or buy machinery, but the rates of operation are very low. So, they can't pay unless it's cheap, the cheapest in the world. Three times cheaper than in China, even with the exchange factor.

Masahiro Sakane, Councilor, Komatsu Ltd.

1963   Joined Komatsu Ltd. after graduation from Osaka City University in
        Engineering
1989   Director
1990   COO of Komatsu Dresser Company (Currently Komatsu America Corp.)
1999   Executive Vice President
2001   President
2007   Chairman & Representative Director of the Board
2010   Chairman of the Board
2013   Director and Councilor
2013   Councilor

Komiyama I see.

Sakane There are plenty of customers, so we sell a lot in Japan, but we don't make money. To put it very bluntly, in order to survive in Japan, we are selling machinery as a service. On the other hand, for giant mines in Chile and Australia, Komatsu develops transport systems with driverless dump trucks, and they are a big factor in improving the mine operations. They bring significant added value for the customer, in safety as well as productivity, and the returns for us are significant too. In order to bring the solutions that work for that type of worksite into construction sites in general, our first step was the SMARTCONSTRUCTION initiative that began in Japan in 2015. You see, the work at almost any civil engineering site is managed with two-dimensional drawings, which means the terrain is not accurately understood. Komatsu built capability for 3-D surveys, using drones. With the depth dimension added, we rolled out smart, semi-automated machinery for general construction use. That alone immensely speeded up the completion of construction jobs. Now we've moved on to issues like efficiently scheduling the arrival of the trucks that carry out the dirt, using our expertise in productivity control to support job execution planning. By now we've brought SMARTCONSTRUCTION to more than a thousand construction sites in Japan. If this pace continues, in the future there will be fewer machines and fewer workers in Japan than there are today. However, the productivity of the nation as a whole will increase.

Komiyama By how much?

Sakane For example, 3-D surveying with drones is overwhelmingly the most efficient method, starting with the fact that it requires fewer workers and fewer man-hours than traditional 2-D surveying. Plus, the subcontracting structure that permeates the construction industry becomes irrelevant. When the internet of things arrives at the construction site, productivity goes up. The contractor and the job performer become one, which is most efficient. This changes the structure of the construction industry. It cuts right to the problem that every industry in Japan is facing, which is a war of attrition among a large number of players. By my lights, with 120 million people on our small islands, a labor shortage is something that doesn't make sense. The natural way is for someone retired from one business to go into some other business, but that sort of reform hasn't happened in Japan, and so we call it labor shortages.

At Komatsu, we carried out a major structural reform, trimmed the workforce, pulled out from businesses where we can't try to be number one or two worldwide, and now we are moving forward with new business that strongly utilizes the internet of things. That is how to overcome deflation. No matter how much monetary easing there is from the government and the Bank of Japan, we can't overcome deflation unless the private sector takes an active role through those kinds of initiatives.

Komiyama We often hear that progress in artificial intelligence and information technologies means the loss of jobs. How do you see it?

Sakane SMARTCONSTRUCTION is the perfect example for that as well. As I mentioned, it means fewer surveyors are required than in the past. However, after the 3-D images are shot, much better data will be obtained if they are read by people with surveying experience. So, losing a job shouldn't mean there is no further work. And there's a good chance that the change of work will create fresh added value.

Komiyama That is clearly the best scenario. I wonder if it will really happen.

Sakane It will, if you ask me. Japan has a problem of companies with non-productive employment, so there is no true workforce shortage, right? And change is the only way, right?

Japanese Factories Are Competitive

Komiyama What did you have to say about regional revitalization?

Sakane Our company name comes from the town of Komatsu in Ishikawa Prefecture. We are one of just a few big companies named for the place of origin.

Komiyama That is true.

Sakane Coming back to Komatsu as a microcosm, in the early postwar years our head office was still in Ishikawa Prefecture, as were most of our factories. Later the head office moved to Tokyo, and as our market became centered along the booming Pacific Coast, the Ishikawa factory locations became inconvenient and we opened factories in Osaka and the Tokyo area. By then the strong yen had arrived, and we plunged into offshore production. So, we went from Ishikawa to the Pacific Coast, and then offshore.

Komiyama Indeed.

Sakane Then in 2001, when I became president, we reassessed whether Japan's competitiveness was really so low. We found that with all the businesses and product lines we were engaged in, the indirect operations in our headquarters divisions, IT departments, etc. were where there was waste, and meanwhile we were doing well in terms of pure monozukuri (making things). In fact, looking only at the variable costs, at the time the exchange rate where Japan and the United States would be on equal terms was about 72–73 yen to the dollar. Since we weren't losing on variable costs, we took a fresh look at our Japan production and investment strategy.

Komiyama You say 72 or 73 yen, that must have been Komatsu only.

Sakane Between you and me, at that time, when the yen was at 110 to 120 to the dollar, other companies too would surely have shown a competitive edge in comparisons of variable costs alone. But Japanese companies generally do not choose to break out their variable costs. Which is to say that executives are wedded to the idea that onsite employment is an untouchable fixed cost. The reality is it is a variable cost. When we take it as a variable cost, in the case of Komatsu, Japan right now is slightly behind Thailand and China, but definitely not behind the US or Europe. Japanese monozukuri skill provide a competitive edge, and therefore once again we come around to Japan. These past 5 years, all the new Komatsu factories were built in Japan.

Komiyama Where are they?

Sakane In places like Kanazawa, Ishikawa and Hitachinaka, Ibaraki – near seaports, because they produce for direct export. From Kanazawa, the machinery is usually carried to the Port of Busan in Korea for onward export to the rest of the world.

Komiyama Busan has become quite a hub.

Sakane Japan accounts for about 20% of our two trillion-yen sales turnover, and about 50% of our production.

Komiyama Interesting.

Sakane That is why I say Komatsu is a microcosm for regional revitalization. It's not that we lost confidence and left Japan, but that we regained confidence in the strength of Japan, to the extent of fresh investment here. But the Japan market proportion could go from 20 to 10% if deflation keeps on at the present rate. I let them know that there is indeed a point where we wouldn't be able to maintain production scale in Japan, and so in any event, we simply have to overcome deflation.

Relocation of Some Head Office Functions, 3.2 Times More Children

Komiyama Some head office functions were moved to Komatsu in 2002.

Sakane At the time, there were roughly 1200 people in the head office, and it was becoming apparent that some of the departments here would be better located out in the field. We began by moving the purchasing department to a factory, which made more sense. Then in 2011 the education groups that had been scattered across Japan were consolidated in a new training center, the Komatsu Way Global Institute. It's located in Ishikawa near the Komatsu Airport, which has direct connections to Narita as well as Incheon International in Korea, so there is no problem at all for trainees coming from anywhere in the world.

Komiyama I see.

Sakane Each year a total of 30,000 people attend the training institute. For their lodging and meals, we use many local catering and lodging services, constantly checking their availability. Each year, to sponsor those 30,000 trainees, we put 700 million yen into the local economy.

Komiyama Very impressive.

Sakane After the Tohoku earthquake in 2011, we decided to direct the capital investment for our Japan factories into all-out energy conservation programs. Many of the Komatsu factories in Japan date from the high-growth period around the 1960s, so about half of the buildings were more than 40 years old. After the Awazu Plant was remodeled in 2014, the new assembly factory used 90% less utility power per unit of production.

Komiyama How did you manage to reduce utility power 90%?

Sakane The big breakthrough was tapping into groundwater. We held a brain-storming session on ways to lower power consumption, and one of our employees said, "I don't see why we're talking about solar and biomass and wind. We're in Ishikawa, which has fabulous groundwater resources." The temperature of that groundwater is 17 °C.

Komiyama Really? So, there's no need for cooling.

Sakane But there wasn't enough energy for heating, so we turned to biomass. The region has an excess of thinned wood, which we started using to generate electricity and provide heating. All the production equipment was designed for energy efficiency, plus the productivity per unit area was doubled, and we also utilized renewables including solar, biomass and wind, so in the end we cut the utility power use by 90%. We are now renovating buildings at the other Japan factories, one by one.

Komiyama I know that one of the reasons behind the decision to relocate some head office functions was the data that had come out on fertility rates among your female employees.

Sakane According to the most recent data for married women over 30, in Tokyo the average number of children is 0.9, while in Ishikawa it's 1.9, and for Ishikawa women in management positions, the average is 2.6 children. Also, the proportion of married women in Tokyo is about 50%, versus about 80% in Ishikawa. When those figures are combined, it works out to 3.2 times more children.

Komiyama That's just wonderful.

Sakane I've met with women managers. They told me they want to have a lot of children and still carry on with their careers. One of them said, maybe half joking, that she has a three-generation household and her father-in-law tells her, "I'll watch the grandkids, you go to work,' so she's able to work very happily.

Komiyama The other side is that in Tokyo, when a woman becomes a full-time housewife, she tends to stay home all the time with her child. It has been found that full-time housewives are much more likely to report that they have felt an urge to abuse their child. It's important to have a setting where the mother can raise children while working.

Sakane There are people who prefer not to have children, and people who want to but can't, so it's impossible to generalize. I myself have three grandchildren. I am 75 now, and I'm sure I would be sad if I didn't have grandchildren at my age. That makes me feel that a three-generation household is the happiest option for seniors. In Komatsu, more than 20% of the people have three generations of their family living within the city, if not in the same home. I think that's part of the goodness of life in local communities.

Komiyama Weren't there some employees who were against the idea of relocating head office functions?

Sakane Indeed, there were. Including my wife. She saw a story in a local newspaper that said Komatsu was relocating its entire head office, and she came to me and said, "What's this? I'm not moving, so you'll be going there alone!" So, when we relocated the purchasing department, we asked about 30 people to give it a try for 3 years, promising that within that time, if they preferred, we could find local people to take their jobs and we would welcome them back in Tokyo. In the end, some decided to buy homes and move their families to Komatsu, and surprisingly few returned to Tokyo.

Retirees Teaching Science to Young Children

Sakane In order to maintain and expand our base in Ishikawa, we need to steadily take on young employees. It turns out that in the Ishikawa area, about one in ten persons works part-time in agriculture. We realized we had to give them some support, and in 2013 we began offering technical assistance for crop and tree farmers. The obvious thing was the intelligent Machine Control Bulldozer from our SMARTCONSTRUCTION line. When our people talked to farmers and asked why the quality of their rice was uneven, they said a completely level rice paddy will produce the most consistent crop, and so we leveled their land using intelligent Machine Control Bulldozer. They were told that in a level field they didn't have to plant rice seedlings, they could just sow the seeds directly. They gave that a try, and their rice crop last year was quite delicious. Now we are spreading the direct sowing method to more farms.

Komiyama Doesn't that mean lower productivity?

Sakane It's actually up by 20%. It's not the traditional furrowed rice paddy, so even young people with no experience can do the work. This is coming into use all over Japan.

My point is that the circle of local energy is gradually expanding. With renewed confidence in our competitiveness, our return to the Ishikawa area we once had fled was a springboard. We brought energy efficiency and biomass fuel into the factory. And once we started the assistance for farming and forestry operations, the agricultural cooperative and the local government and the banks stepped up with financing aimed at increasing productivity.

The training institute also deserves to be used by more than just our own people. We thought of setting up science classrooms for young children, and as a centerpiece we brought one of our mining trucks back from Chile, the world's largest dump truck with tires four meters in diameter and a price tag of half a billion yen, and we placed it at the training institute. Now we have about 50,000 adults and children visiting each year. We run classrooms on science and monozukuri, with themes like what makes electricity occur, or what are the principles behind the transport of heavy objects. The people teaching this material are our retired employees.

Komiyama Fascinating.

Sakane Each year about 300 people take shifts teaching the children.

Komiyama Are they paid for their work?

Sakane We pay a little by the day, and it's gradually going up. And what the teachers tell us is they find they need less medical care.

Komiyama What a great story. How about linking up with the local schools?

Sakane We're already working with the primary schools in the city of Komatsu. Every fifth-year student comes on a field trip to the training institute.

Komiyama Japan has various challenges in education, including science learning and English learning, and especially the problem of bullying. Primary school teachers are very capable, but they often have limited life experience. People who have retired from companies, on the other hand, have a lifetime of experience, perhaps including work assignments outside Japan. Bringing their assistance into the classroom could lead to big changes in our schools.

Sakane You're right. There are quite a few of our former factory employees in the area. These are the kinds of people whose service in education and public administration is indispensable for revitalizing local communities.

Komiyama In our aging society, naturally there is a lot of discussion about caregiving. But it's also important to consider how active our seniors are. For example, the Mayekawa Manufacturing Company has a series of very successful niche products that were developed mainly through collaborations of their former and current employees. People who are working may be too busy, and drawing on the capacities of our retirees who are blessed with free time will be a real key for growth.

Interview 3: Considering Ways to Solve Social Problems

There Is Much Room for Innovation of the Social System

**Interview with Hiroshi Yoshikawa, Professor of Economics at Rissho University
By: Ryuji Konishi, Visiting Professor, Ritsumeikan Asia Pacific University
  Graduate School**

Konishi This book was written in regard to how to realize a Platinum Society from the viewpoint of science and technology. Human society has made progress, but various new problems, such as global warming, have appeared. However, such problems can be solved by science and technology. The basic structure of the logic here is that if we can change the social mindset of how to use science and technology as it advances, an affluent circulating society would emerge in due course within a couple of decades.

However, social structure, social systems, etc., are not adequately addressed here. Basically, it really needs to engage with the issue of how we should understand things that are not directly influenced by science and technology in order to secure the Platinum Society. I requested this interview with you, Professor Yoshikawa, because if I think that we must find a way to solve the latent problems related to social security, health care and the like due to an aging population if Japan were a leading country in resolving societal problems. Today I would like to ask you to discuss how we should think about the relationship between taxation and social security, as well as the framework of a market economy and the state.

Hiroshi Yoshikawa, Professor, Faculty of Economics, Rissho University

1974   B.A. in Economics, The University of Tokyo
1978   Ph.D. in Economics, Yale University
1978   Associate Professor, State University of New York, Albany
1982   Associate Professor, Institute of Social and Economic Research, Osaka
        University
1988   Associate Professor, Faculty of Economics, University of Tokyo
1993   Professor, Faculty of Economics, University of Tokyo
2016   Professor, Faculty of Economics, Rissho University

The Notion that a Decline in the Population Means the Economy will Falter Is Faulty

Yoshikawa I recently wrote a book called *Demography and the Japanese Economy*. What I say here is that, well, the fact that the Japanese population is declining is well known. According to the National Institute of Population and Social Security Research, the moderate-range estimate of birth rates about 100 years from now is projected at around, I guess, 40 million. This prediction means that the population of 120 million will become a third of that size. I also think that such a sudden drop in the population is a problem. However, it is slightly off the mark to characterize it as an economic problem. My basic idea is that today's pessimistic theories that are popular, like that present-day Japan will only continue to decline because of the population decrease, that the Japanese market is finished, or that the Japanese economy is over, are all wrong. What will be key is innovation. In that regard, my idea is exactly the same as the so-called Platinum Society idea of yours, Professor Komiyama.

This is just an aside, but about 2 years ago, I participated in a German-Japanese conference on economics in Berlin. My visit then left a strong impression on me. As you know, Germany is also a country undergoing a severe population decline, and they recognized that as a problem. Their solution to the problem of a declining population was immigration, but leaving that specific approach aside, there was absolutely no pessimistic theorizing that the German economy will falter and decline from here on out. The basic atmosphere was that the German economy would continue to be strong into the future.

To introduce some numbers, it is well known that in real terms, the Japanese economy grew by approximately 10% during Japan's economic boom, which is approximately from 1955 to the early 1970s. However, what most people do not know is the rate of growth in the workforce population. The answer is 1%; to be more precise, the actual figure may have been about more like 1.2%. The difference of 9% between the 10% growth in the economy and the 1% growth in the workforce population accounts for a 9% increase in income per person. As to why income increased, that can basically be attributed to innovation and investment in capital stock. Metaphorically speaking, imagine that bulldozers, cranes, etc. are introduced to a construction site where there only used to be picks and shovels. Just the way you must have technology for that, you also at the same time actually need to have the bulldozers, cranes, etc., introduced at the construction site. That's what I mean.

Ryuji Konishi, Visiting Professor, Graduate School of Management, Ritsumeikan APU

1967	Joined The Long-Term Credit Bank of Japan, Ltd. After graduation from The University of Tokyo in Law
1989	Completed AMP, Harvard Business School
1993–1998	Executive Managing Director & Member of The Board
1999–2000	Lecturer, Graduate School for Asia-Pacific Studies, University of Waseda
2000–2003	Director & Member of The Board, GlaxoSmithKline Japan
2001–2008	Professor, Graduate School of Industrial Management, University of Kyushu

It Is Innovation for Being Used at the Site

Yoshikawa Perhaps this is what's called social engineering. In the case of Japan, even if something is achieved at the level of science or technology, whether or not that achievement gets introduced into society is a separate problem. There is a huge social problem. In that regard, I believe that there is still much room for improvement in Japan. I believe that there is also a large role to be played by the state and the government. For example, a historical example that was successful that I think of, though maybe somewhat out of the blue, is the Dojunkai Apartments. In the recovery after the 1923 Kanto Earthquake, the Home Ministry erected about twenty Dojunkai Apartment buildings in Tokyo and Yokohama. Up until then, collective housing had consisted of wooden buildings with at most two stories; however, the collective housing introduced with the Dojunkai Apartments consisted of concrete structures that were 3 to 4 stories, or, in some cases, even 6–7 stories high. The Dojunkai Apartments served as the model for post-WWII collective housing, and it is clear that they represent the prototype of the modern-day apartment complexes. The Home Ministry played a leading role in carrying out that experimental project.

Because "seeing is believing." it is extremely important to present a model in a form that people can see with their eyes. Especially nowadays, I think it's even more necessary to make a model. This is true especially in the fields of medicine and caregiving. An easy to understand example would be that even if a caregiver robot were developed, it is not introduced into an actual caregiving facility. It is because they are not a part of the medical treatment fee structure. Aside from the case where there is a unique business manager who wants to experiment with the robot and pays out of his/her pocket for it, there is no basis for introducing such robots into caregiving facilities. Thus, the state must play a leading role in spreading the use of caregiver robots. People have been talking about this for 10 years now, but it still isn't happening.

Even in the field of energy, in which you are interested, there was a certain type of scheme that was created in order to foster the diffusion of renewables. In the same manner, the medical and caregiving fields will more than likely require some sort of public initiative before innovative technologies are adopted for widespread use. I guess this is what is called social innovation. I believe that there is a lot of room for progress there.

This is somewhat out of the blue, but there is an example I find interesting in economics that relates to disposable diapers for the elderly. The domestic market for baby diapers has hit the ceiling in the midst of declining birthrates. However, the demand for elderly people is steadily expanding and pulling the industry along. The only difference between baby diapers and diapers for the elderly is the surface area, so there is probably no technical innovation there. This is an example that makes it easy to see that innovation is not limited to hardware engineering, and clearly shows that this other form of innovation is also needed.

Komiyama It is important to combine innovations that are taking place at various levels in many fields.

Disparity Widens Amid the Ongoing Downturn

Yoshikawa I'd like to touch on the relationship between taxation and society raised by you, Professor Konishi. This problem has its origins in the problem of disparity. That is to say, it is a problem related to how many of a given thing to produce, and how to divide it up. In economics jargon, this is referred to as "distribution." With respect to income, it is called "income distribution." In other words, there is only one answer up until the problem of producing as efficiently as possible, but there is more than one answer to the question that addresses fairness when it comes to dividing up the product. That is because there are a lot of value judgments involved.

As you know, several years ago the book, *Capital in the Twenty-First Century* by French economist Thomas Piketty became a best seller worldwide. Piketty was not the first to point this out, but with respect to the limited scope of developed countries, his book made the fact clear that before World War II, developed countries had extremely disparate societies, but that the extremely wealthy class disappeared as a result of the war, and in a word, society became more egalitarian. That situation has changed drastically since around the 1980s. Since then, the United States is a country in which the wealthy have steadily increased their wealth to become super wealthy. The annual income of top management at large listed companies, in terms of multiples of the average income of the employees of said companies, had been about 40 times the average employee's income until about 30 years ago, but has climbed to 400 times over these past 30 years.

In Japan, it is not the case that the top earners in society are steadily becoming super wealthy; however, social disparity has widened. The problem in Japan is that the whole country's economy suddenly toppled. In other words, it is not that the economy bred a group of winners and a group of losers, but to put it in somewhat exaggerated terms, everyone in Japan became a loser. I think that the average to mid-level incomes have fallen sharply. That is to say, at the lower end of the income spectrum, particularly in the middle class, incomes have fallen.

Komiyama In addition, is it that there has not been a rise in income for the upper end of the income spectrum?

Yoshikawa In exaggerated terms, yes. And it is always the strata with a lower level of income that bears the burden in such situations. To speak about it abstractly, there is the problem of regular/non-regular employees, as you know. Thirty years ago, non-regular employees are said to have accounted for about 16% of the workforce, whereas today that figure has reached about 40%.

Komiyama Some sources report that it has exceeded 40%.

Yoshikawa In addition, with regard to Japan, it's well known that disparity is a particularly large problem with respect to the aging population. That is to say, if you took a sample group of one million people in their 20s, there would be differences in income, assets, health, etc., but they wouldn't vary widely. On the other hand, if you took a sample group of one million people in their 70s, there would be a huge degree of variance. With an aging population, the proportion of people that are elderly increases across society; accordingly, it's natural that there will be a correspondingly wider degree of variance across society and the economy as a whole.

Expansion in Social Security Costs Must Be Covered by the Consumption Tax

Yoshikawa The system that to some degree ameliorates the problem of disparity across society is the social security system. It's based on the idea that the state should be held responsible to narrow the disparity with regard to problems such as pension, health insurance, and nursing care systems, where the disparity will widen too much if neglected.

Here, the question is raised as to how to finance those systems. At present, social security benefits amount to about 110 trillion yen. Sixty percent of that is covered by health insurance fees. If 100 is used to represent the total expenditures on social security benefits, as you know, health insurance fees are borne in equal shares by the employer and employees, and the fees cover 60% of the costs; accordingly, the remaining 40% represents a deficit. The national government can cover 30% that 40%, while local governments can cover the 10%. The burden of the national government is approximately 30 trillion yen, and that can be appropriated from the general account budget.

Government expenditures for public works, education/science and technology, ODA (government provided overseas development aid), defense, etc., are fundamentally at a zero growth rate. The public policy expenditure that will grow on its own is that related to social security. Under normal circumstances, the government spending on social security is increasing at about 60–70 billion yen per year. Accordingly, social security expenditures are the cause of expansion in the government budget deficit. That is not the case only in Japan, but in the United States and Europe as well. The situation in Japan is the most extreme, though.

So, what can we do? Simply put, Japanese people have to pay a bit more taxes. Taxation can become long topic of conversation, so to be brief, consumption tax does not have a progressive structure like income tax, but it is a proportional tax, so is not a bad idea. It may be necessary to provide some allowance to low-income people due to a regressive effect, but consumption tax is the answer.

In Europe, EU policy stipulates a minimum taxation rate, which for consumption tax is 15%. However, there are almost no countries in the EU with the minimum consumption tax rate of 15%: in England, France, and Germany, the rate is approximately 20%; and in Sweden and Norway, it is approximately 25%. There is no way any country in the EU could have a consumption tax rate of 5%, which was the rate in Japan until a short while ago. Accordingly, there was impetus to do something regarding the consumption tax, which should have been raised to 10% but was raised to 8% and has remained there, with calls to put off raising it further into the future.

Komiyama Yes, I certainly understand that it is necessary to finance social security through indirect taxation. Meanwhile, I believe that it is also a fact that the economy is not growing because there is not enough innovation happening. By implementing the concept of "The Platinum Society" we have put forth, a plethora of opportunities for business will be created through pursuing efforts aimed at improving the Quality of Life (QOL).

Yoshikawa If you ask me, I would say that it is innovation that will help solve the problem of an aging population, environmental problems, etc. That is because the solutions don't lie in how many of an existing product is sold, but in creating new products. The size of the Japanese market is ideal for conducting various experiments. To borrow your words, it is as you say that leading country in resolving societal problems has more opportunities for business commensurate with their status.

Komiyama The Japanese system and its mentality are impeding such pursuits. For example, new business models such as ride sharing like Uber has a hard time becoming popular in Japan.

Yoshikawa This is a cliché, but don't you think there is a lack of an enterprising spirit? Going back to the story of disposable diapers, I happened to hear from the person who developed this. This person told me that the proposal was finally accepted on the third try. We can also add to that why the first robot vacuum cleaner was not developed in Japan. Apparently, it had been developed, but there were negative opinions like "what were they going to do if the robot bumped in to the alter, knocking over the candles, and starting a fire," so they couldn't launch it in the market.

The Government Should Play the Role of Coordinator

Konishi Earlier, we discussed the involvement of the national government in relation to some of these issues. It bears noting that during the period of the economic boom, the then Ministry of International Trade and Industry played a proactive role, setting targets and cultivating industries. However, we have reached a stage where it is not even clear as to what extent the state should play. The national government doesn't even have a policy for tackling the problem of non-regular employees. What type of relationship should there be between the state and the market in the future?

Yoshikawa I actually think that is not something we can discuss in general terms. However, in the medical and caregiving fields, there are no market-making players other than the government. More specifically, there are engineers here and there developing caregiver robots. Even so, private organizations are not capable of having such robots introduced at care facilities for experimental uses and the like, or having them incorporated into the caregiving insurance system so as to promote their use.

Komiyama You are absolutely right. For example, in order to make use of abandoned arable land, Ito En is establishing tea farms, but this has only become possible now that businesses are allowed to enter in agricultural enterprises. It is the national government that creates such systems. However, in the forestry industry, because the productivity per unit area is not high, private enterprises cannot easily undertake such ventures. It may be that there is a necessity to reorganize the delineation with respect to how far the state should be involved and from where private enterprises should be left to their own devices.

Yoshikawa It varies from industry to industry, but I would like to see the government doing everything it should.

Komiyama Renewable energy has seen growth because the government created a system whereby they buy renewables at a fixed price.

Yoshikawa There are some people who claim that it created a bubble, but if you look at it holistically, I believe this is just a part of the ups and downs. I think that there is a great role to be played by the country and the government.

Komiyama I also think so.

Yoshikawa The government must, I believe, play a role as a coordinator. That is to say, responding to the problems related to the environment and energy, and the aging society is certainly necessary for humanity in the twenty-first century. It seems to me that, as opposed to having a single industry solve a given problem, such as the aging society, for example, that the entire system will need to be revamped. It seems that everything, like buildings, transportation, and distribution – almost everything as we know it – will change.

Konishi Yes, the industrial structure changing to that of the Platinum Society.

Yoshikawa The technology developed in each specialized field has to be bundled together and integrated horizontally, so that what are essential should be produced at the end.

Komiyama I believe that we are entering an era where that will be possible. If we can appropriately combine technologies and knowledge we already have, it can be done. That's why the horizontal axis becomes important.

Yoshikawa For example, the vehicles used in the emergency medical field will probably become smart cars, but that means that the departments that become the destination points for those emergency vehicles will also have to incorporate new technology and change. This is where it becomes necessary that there be a coordinator for facilitating the horizontal integration.

Komiyama If an attempt were made to configure the emergency medical system by new technology such as drones, GPSs, smart cars, etc., it would run up against the vertical hierarchy of the government bureaucracy. So, I believe that this is the type of problem that needs to be solved by the government.

Acknowledgements

We would like to take this opportunity to express our heartfelt appreciation to all those who have been involved in this project. Much of the genesis of the content of this book derives from observations and insights garnered through discussions with a large number of contributors.

In particular, we wish to extend our sincere gratitude to the following, for their support and cooperation during the writing of this book:

Corporate members (CEOs), local government members (mayors and Governors), and special members (academia) of The Platinum Society Network;

Representatives of the Presidential Endowed Chair for "Platinum Society", the University of Tokyo;

Representatives of the Center for Low Carbon Society Strategy (LCS) at the Japan Science and Technology Agency (JST) (National Research and Development Agency);

Members of the 2050 Technology and Management Knowledge Development Forum;

Representatives of the Platinum Society Research Association at the Mitsubishi Research Institute;

Representatives of the Executive Management Program (Todai EMP) at the University of Tokyo;
and
Representatives of the Association of Digital Textbook and Teaching (DiTT).

We are also indebted to the people whose names are written below as they have directly contributed to the making of this book by providing interviews, manuscript drafts, data, photographs, materials, and so on. Our utmost thanks go to all of you for your invaluable cooperation and assistance.

Takeshi Uchiyamada
Masahiro Sakane

Hiroshi Yoshikawa
Ryuji Konishi
Kazuhiko Ishimura
Kazuhiko Takeuchi
Masao Maekawa
Kuniaki Kawamura
Hachiro Honjo
Hidemitsu Sasaya
Hirosaku Nagano
Motohiko Nishimura
Junichi Fujino
Toshiharu Ikaga
Ryozo Ooka
Frank Peter Popoff
Takeshi Kuroda
Mitsuaki Hayashi
Suguru Noda
Toshio Osawa
Fumio Ohue
Satoshi Koike
Minoru Yamamoto
Masako Konishi
Tatsuya Okubo
Yasunori Kikuchi
Nanako Ishido
Takayuki Kamikura
Shigehiro Sasaoka
Zhanfei (Jonathan) Li
Sugiura Shogo
Toshitake Chino
Jun Yasuki
Hiroko Yamabe
Emi Inuyama
Katsuji Kasajima
Hiroshi Kimura
Tomoo Aota
Kimie Yajima
Hiroki Komazaki
Shigeo Kai
Tomohiro Inoue
Tomoko Iwata
Kanako Tanaka
Rika Itaya
Kyota Omori
Tomoo Matsuda

Toru Higaki
Toru Hashi
Kuniyuki Nishimura
Miho Umeda
Hiroshi Iwase
Naoki Yoshida
Hozuma Sekine
Shinichi Kamei
Akiko Kato
Shuichiro Kishi
Norio Shigetomi
Hisako Takahashi
Kazuki Motohashi
Takahiro Oyama
Mitsuru Irie
Nobuhiro Umehara
Karin Hosumi
Kazunari Suzuki
Ayako Mizushima
Dinh Minh Hung
Nao Ozawa
Shota Yamaguchi
Rie Arai
Takashi Endo
Tetsuya Nomoto
Satoko Horie
Satoshi Waseda
Tsutomu Tanaka
Michelle Poh
Kira Kamilla Smiley
Daniel Walter
Hiroshi Iwasaki
Teruo Mitsumori

Lastly, we would like to express our sincere gratitude to freelance journalist Aiko Hayashi, and Taro Tanaka of Nikkei Business Publications, Inc., for their collaboration vis-à-vis the publication of this book.